OIL WELL TESTING

OIL WELL TESTING

Deepak Sharma

RANDOM

RANDOM PUBLICATIONS

NEW DELHI - 110 002 (INDIA)

Oil Well Testing

ISBN 978-93-51116-69-1

Published in 2015 in India by

RANDOM PUBLICATIONS

4376-A/4B, Gali Murari Lal, Ansari Road
New Delhi-110 002
Phone: +9111-43580356, 23289044
E-mail: randomexports@gmail.com; sales@randompublications.com;
info@randompublications.com

Reprinted 2026

Type Setting by: Friends Media, Delhi-110089

Preface

In the petroleum industry, a well test is the execution of a set of planned data acquisition activities to broaden the knowledge and understanding of hydrocarbons properties and characteristics of the underground reservoir where hydrocarbons are trapped. The test will also provide information about the state of the particular well used to collect data. The overall objective is identifying the reservoir's capacity to produce hydrocarbons, such as oil, natural gas and condensate. Data gathered during the test period includes volumetric flow rate and pressure observed in the selected well. Outcomes of a well test, for instance flow rate data and gas oil ratio data, may support the well allocation process for an ongoing production phase, while other data about the reservoir capabilities will support reservoir management. There are many flavours of well tests and various ways to categorize test types by its objectives, however two main categories only by objectives, these are productivity tests and descriptive tests. According to The Lease Pumper's Handbook of Oklahoma Commission on Marginally Producing Oil and Gas Wells, there are four basic well test types: potential tests, daily tests, productivity tests, and gas oil ratio tests, the latter three in the broader productivity test category. Test objectives will change throughout the different phases of a reservoir or oil field, from the exploration phase of wildcat and appraisal wells, through the field development phase and finally through the production phase, which may also have variations from the initial period of production to improved recovery by the end of the field lifecycle time. Professionals working with reservoir modelling may get information about the rock permeability from core samples. Other sources of information to the model are well log data and seismic data, but such data are complementary only, and for example, seismic data is insufficient to interpret whether a structural trap has been sealed. Information from well tests will supplement the amount of information with flow rate data, pressure data, and other, which is needed to build a rich reservoir model. The main objective in the exploration phase is to assess the size of a reservoir and state with a given certainty whether it has the properties for commercial exploitation and shall contribute to accounting for available reserves.

Well testing taking place before permanent well completion is referred to as drill stem testing. The reservoir model is further developed to support the field development planning and to advise for the optimal location for extra production wells to be drilled. Descriptive well tests are designed and performed in the new wells. The total flow rate of all wells in total are measured, but the contributions of the individual wells are unknown. It is important to know the individual contributions to account hydrocarbon material balance and for well monitoring and reservoir management. To obtain individual well flow rates, it is common to use a smaller test separator. This is an isolated and down-scaled processing system in parallel with the normal flows. Regularly, for example once a month per well, the flow from one and only one selected well is led into the test separator for determining well flow rate for the selected well. The separator divides the flow from the well into the streams of individual products which typically are oil, gas and water, but may include natural-gas condensate. Contamination may also be removed and fluid samples collected. This helps to allocate individual flow rate contributions, but the method has uncertainties. Flow rate, water cut, GOR and other parameters for the test system can deviate from production separators. This is generally taken into account by the allocation of products back to individual wells based on the field total, and by using data from the individual well tests.

This valuable book covers all the major operational aspects of oil and gas well testing and uses a structured approach to guide the reader through the steps required to safely and effectively plan a well test operation.

I thank all members of my team who have helped in the preparation of the book. My special thanks go to "Random Publications" who have published the book.

— Deepak Sharma

Contents

Chapter 1
Introduction

Well Test

A well test is conducted to evaluate the amount of water that can be pumped from a particular water well. More specifically, a well test will allow prediction of the maximum rate at which water can be pumped from a well, and the distance that the water level in the well will fall for a given pumping rate and duration of pumping.

Well testing differs from aquifer testing in that the behaviour of the well is primarily of concern in the former, while the characteristics of the aquifer (the geological formation or unit that supplies water to the well) are quantified in the latter.

When water is pumped from a well the water level in the well falls. This fall is called drawdown. The amount of water that can be pumped is limited by the drawdown produced. Typically, drawdown also increases with the length of time that the pumping continues.

Well Losses vs. Aquifer Losses

The components of observed drawdown in a pumping well were first described by Jacob (1947), and the test was refined independently by Hantush (1964) and Bierschenk (1963) as consisting of two related components,

$$s = BQ + CQ^2,$$

where s is drawdown (units of length e.g., m), Q is the pumping rate (units of volume flowrate e.g., m^3/day), B is the aquifer loss coefficient (which increases with time — as predicted by the Theis solution) and C is the well loss coefficient (which is constant for a given flow rate).

The first term of the equation (BQ) describes the linear component of the drawdown; i.e., the part in which doubling the pumping rate doubles the drawdown.

The second term (CQ^2) describes what is often called the 'well losses'; the non-linear component of the drawdown. To quantify this it is necessary to pump the well at several different flow rates (commonly called *steps*). Rorabaugh (1953) added to this analysis by making the exponent an arbitrary power (usually between 1.5 and 3.5).

To analyze this equation, both sides are divided by the discharge rate (Q), leaving s/Q on the left side, which is commonly referred to as *specific drawdown.* The right hand side of the equation becomes that of a straight line. Plotting the specific drawdown after a set amount of time (Δt) since the beginning of each step of the test (since drawdown will continue to increase with time) versus pumping rate should produce a straight line.

$$\frac{s}{Q} = B + CQ$$

Fitting a straight line through the observed data, the slope of the best fit line will be C (well losses) and the intercept of this line with $Q = 0$ will be B(aquifer losses). This process is fitting an idealized model to real world data, and seeing what parameters in the model make it fit reality best. The assumption is then made that these fitted parameters best represent reality (given the assumptions that went into the model are true). The relationship above is for fully penetrating wells in confined aquifers (the same assumptions used in the Theis solution for determining aquifer characteristics in an aquifer test).

Well Efficiency

Often the *well efficiency* is determined from this sort of test, this is a percentage indicating the fraction of total observed drawdown in a pumping well which is due to aquifer losses (as opposed to being due to flow through the well screen and inside the borehole). A perfectly efficient well, with perfect well screen and where the water flows inside the well in a frictionless manner would have 100% efficiency. Unfortunately well efficiency is hard to compare between wells because it depends on the characteristics of the aquifer too (the same amount of well losses compared to a more transmissive aquifer would give a lower efficiency).

Specific Capacity

Specific capacity is a quantity that which a water well can produce per unit of drawdown. It is normally obtained from a step drawdown test. Specific capacity is expressed as:

$$S_c = \frac{Q}{h_0 - h}$$

where

S_c is the specific capacity ($[L^2T^{-1}]$; m²/day or USgal/day/ft)

Q is the pumping rate ($[L^3T^{-1}]$; m³/day or USgal/day), and

$h_0 - h$ is the drawdown ([L]; m or ft)

The specific capacity of a well is also a function of the pumping rate it is determined at. Due to non-linear well losses the specific capacity will decrease with higher pumping rates. This complication makes the absolute value of specific capacity of little use; though it is useful for comparing the efficiency of the same well through time.

Reservoir Modelling

In the oil and gas industry, reservoir modelling involves the construction of a computer model of a petroleum reservoir, for the purposes of improving estimation of reserves and making decisions regarding the development of the field.

A reservoir model represents the physical space of the reservoir by an array of discrete cells, delineated by a grid which may be regular or irregular. The array of cells is usually three dimensional, although 1D and 2D models are sometimes used. Values for attributes such as porosity, permeability and water saturation are associated with each cell. The value of each attribute is implicitly deemed to apply uniformly throughout the volume of the reservoir represented by the cell.

Types of Reservoir Model

Reservoir models typically fall into two categories:

- Geological models are created by geologists and geophysicists and aim to provide a static description of the reservoir, prior to production.
- Reservoir simulation models are created by reservoir engineers and use finite difference methods to simulate the flow of fluids within the reservoir, over its production lifetime.

Sometimes a single "shared earth model" is used for both purposes. More commonly, a geological model is constructed at a relatively high (fine) resolution. A coarser grid for the reservoir simulation model is constructed, with perhaps two orders of magnitude fewer cells. Effective values of attributes for the simulation model are then derived from the geological model by a process of "upscaling".

Alternatively, if no geological model exists, the attribute values for a simulation model may be determined by a process of sampling geological maps.

Uncertainty in the true values of the reservoir properties is sometimes investigated by constructing several different realizations of the sets of attribute values. The behaviour of the resulting simulation models can then indicate the associated level of economic uncertainty.

The phrase "reservoir characterization" is sometimes used to refer to reservoir modelling activities up to the point when a simulation model is ready to simulate the flow of fluids. Commercially available software is used in the construction, simulation and analysis of the reservoir models.

Geologic Modelling

Geologic modelling or Geomodelling is the applied science of creating computerized representations of portions of the Earth's crust based on geophysical and geological observations made on and below the Earth surface. A Geomodel is the numerical equivalent of a three-dimensional geological map complemented by a description of physical quantities in the domain of interest. Geomodelling is related to the concept of Shared Earth Model which is a pluridisciplinary, interoperable and updatable knowledge base about the subsurface.

Geomodelling is commonly used for managing natural resources and natural hazards and quantifying geological processes, with main applications to oil and gas fields, groundwater aquifers and ore deposits. For example, in the oil and gas industry, realistic geologic models are required as input to reservoir simulator programmes, which predict the behaviour of the rocks under various hydrocarbon recovery scenarios. An actual reservoir can only be developed and produced once, and mistakes can be tragic and wasteful. Using geological models and reservoir simulation allows reservoir engineers to identify which recovery options offer the safest and most economic, efficient, and effective development plan for a particular reservoir.

Geologic modelling is a relatively recent subdiscipline of geology which integrates structural geology, sedimentology, stratigraphy, paleoclimatology, and diagenesis;

In 2 dimensions a geologic formation or unit is represented by a polygon, which can be bounded by faults, unconformities or by its lateral extent, or crop. In geological models a geological unit is bounded by 3-dimensional triangulated or gridded surfaces. The equivalent to the mapped polygon is the fully enclosed geological unit, using a triangulated mesh. For the purpose of property or fluid modelling these volumes can be separated further into an array of cells, often referred to as voxels (volumetric elements). These 3D grids are the equivalent to 2D grids used to express properties of single surfaces.

Geomodelling generally involves the following steps:

1. Preliminary analysis of geological context of the domain of study.
2. Interpretation of available data and observations as point sets or polygonal lines.
3. Construction of a structural model describing the main rock boundaries (horizons, unconformities, intrusions, faults)
4. Definition of a three-dimensional mesh honouring the structural model to support volumetric representation of heterogeneity and solving the Partial Differential Equations which govern physical processes in the subsurface (e.g. seismic wave propagation, fluid transport in porous media).

Geologic Modelling Components

Structural framework: Incorporating the spatial positions of the major boundaries of the formations, including the effects of faulting, folding, and erosion (unconformities). The major stratigraphic divisions are further subdivided into layers of cells with differing geometries with relation to the bounding surfaces (parallel to top, parallel to base, proportional). Maximum cell dimensions are dictated by the minimum sizes of the features to be resolved (everyday example: On a digital map of a city, the location of a city park might be adequately resolved by one big green pixel, but to define the locations of the basketball court, the baseball field, and the pool, much smaller pixels - higher resolution - need to be used).

Rock type

Each cell in the model is assigned a rock type. In a coastal clastic environment, these might be beach sand, high water energy marine upper shoreface sand, intermediate water energy marine lower shoreface sand, and deeper low energy marine silt and shale. The distribution of these rock types within the model is controlled by several methods, including map boundary polygons, rock type probability maps, or statistically emplaced based on sufficiently closely spaced well data.

Reservoir Quality

Reservoir quality parameters almost always include porosity and permeability, but may include measures of clay content, cementation factors, and other factors that affect the storage and deliverability of fluids contained in the pores of those rocks. Geostatistical techniques are most often used to populate the cells with porosity and permeability values that are appropriate for the rock type of each cell.

Fluid Saturation

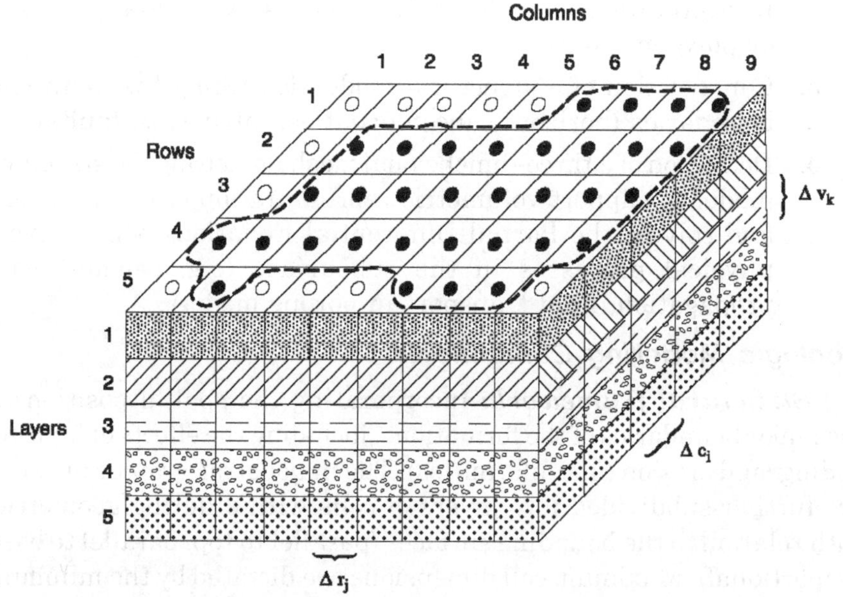

***Figure:** A 3D finite difference grid used in MODFLOW for simulating groundwater flow in an aquifer.*

Most rock is completely saturated with groundwater. Sometimes, under the right conditions, some of the pore space in the rock is occupied by other liquids or gases. In the energy industry, oil and natural gas are the fluids most commonly being modelled. The preferred methods for calculating hydrocarbon saturations in a geologic model incorporate an estimate of pore throat size, the densities of the fluids, and the height of the cell above the water contact, since these factors exert the strongest influence on capillary action, which ultimately controls fluid saturations.

Geostatistics

An important part of geologic modelling is related to geostatistics. In order to represent the observed data, often not on regular grids, we have to use certain interpolation techniques. The most widely used technique is kriging which uses the spatial correlation among data and intends to construct the intepolation via semi-variograms.

To reproduce more realistics spatial variability and help assessing spatial uncertainty between data, geostatistical simulation is often used, based on variograms, training images or parametric geological objects.

Mineral Deposits

Geologists involved in mining and mineral exploration use geologic modelling to determine the geometry and placement of mineral deposits in the subsurface of the earth. Geologic models help define the volume and concentration of minerals, to which economic constraints are applied to determine the economic value of the mineralization. Mineral deposits that are deemed to be economic may be developed into a mine.

Technology

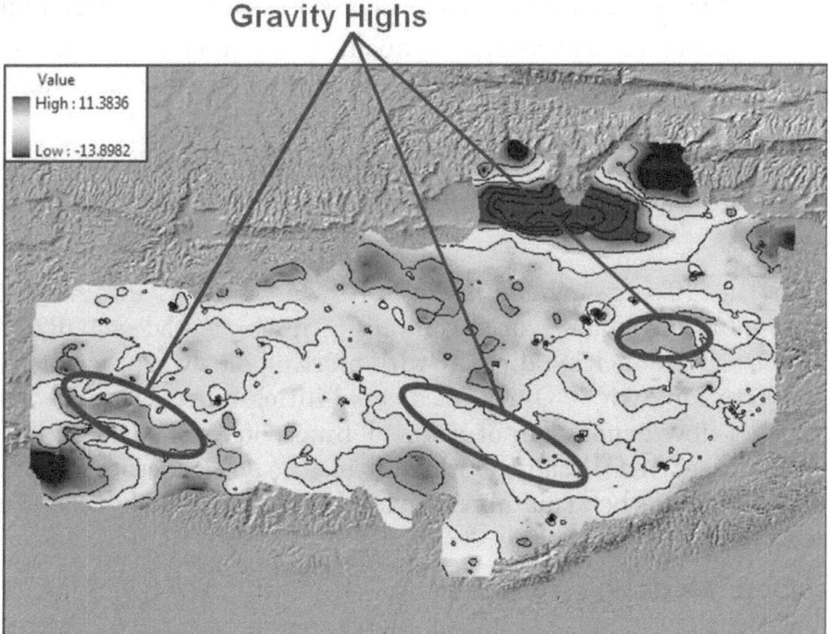

Figure: *Gravity Highs*

Geomodelling and CAD share a lot of common technologies. Software is usually implemented using object-oriented programming technologies in C++, Java or C# on one or multiple computer platforms.

The graphical user interface generally consists of one or several 3D and 2D graphics windows to visualize spatial data, interpretations and modelling output. Such visualization is generally achieved by exploiting graphics hardware. User interaction is mostly performed through mouse and keyboard, although 3D pointing devices and immersive environments may be used in some specific cases.

Geometric objects are represented with parameteric curves and surfaces or discrete models such as polygonal meshes.

Research in Geomodelling

Problems pertainting to Geomodelling cover:

- Defining an appropriate Ontology to describe geological objects at various scales of interest
- Integrating diverse types of observations into 3D geomodels: geological mapping data, borehole data and interpretations, seismic images and interpretations, potential field data, well test data, etc.
- Better accounting for geological processes during model building
- Characterizing uncertainty about the geomodels to help assess risk. Therefore, Geomodelling has a close connection to Geostatistics and Inverse problem theory.
- Applying of the recent developed Multiple Point Geostatistical Simulations (MPS) for integrating different data sources

History

In the 70's, geomodelling mainly consisted of automatic 2D cartographic techniques such as contouring, implemented as FORTRAN routines communicating directly with plotting hardware. The advent of workstations with 3D graphics capabilities during the 80's gave birth to a new generation of geomodelling software with graphical user interface which became mature during the 90's. Since its inception, geomodelling has been mainly motivated and supported by oil and gas industry.

Geologic Modelling Software

Software developers have built several packages for geologic modelling purposes. Such software can display, edit, digitise and automatically calculate the parameters required by engineers, geologists and surveyors. Current software is mainly developed and commercialized by oil and gas or mining industry software vendors:

Packages include:

- JewelSuite by Baker Hughes
- Schlumberger Petrel
- Paradigm Gocad and SKUA
- Geocap
- Landmark Graphics Corporation DecisionSpace Earth Modelling
- Roxar IRAP RMS Suite
- Dynamic Graphics Inc. EarthVision
- Geomodeller3D, Intrepid Geophysics
- GM-SYS, Geosoft
- VOXI, Geosoft
- Gocad
- GSI3D
- FastTracker (Reservoir Modelling)
- ArcGIS
- ARANZ Geo Ltd Leapfrog
- Gemcom Software International Surpac, GEMS and Minex
- Promine
- Micromine

Groundwater modelling:

- FEFLOW
- FEHM
- MODFLOW
- GMS
- Visual MODFLOW
- ZOOMQ3D

Moreover, industry Consortia or companies are specifically working at improving standardization and interoperability of earth science databases and geomodelling software:

- Standardization: GeoSciML by the Commission for the Management and Application of Geoscience Information, of the International Union of Geological Sciences.
- Standardization: RESQML(tm) by Energistics
- Interoperability: OpenSpirit, by TIBCO(r)

Reservoir Simulation

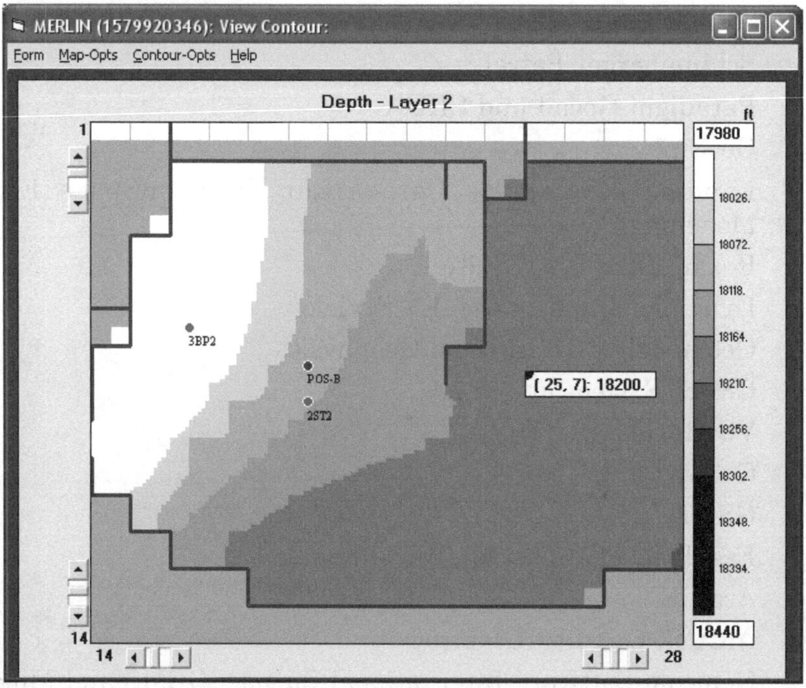

Figure: *A simulated depth map of the geology in a full field model from the Merlin finite difference simulator*

Reservoir simulation is an area of reservoir engineering in which computer models are used to predict the flow of fluids (typically, oil, water, and gas) through porous media.

Uses

Reservoir simulation models are used by oil and gas companies in the development of new fields. Also, models are used in developed fields where production forecasts are needed to help make investment decisions. As building and maintaining a robust, reliable model of a field is often time-consuming and expensive, models are typically only constructed where large investment decisions are at stake. Improvements in simulation software have lowered the time to develop a model. Also, models can be run on personal computers rather than more expensive workstations.

For new fields, models may help development by identifying the number of wells required, the optimal completion of wells, the present and future needs for artificial lift, and the expected production of oil, water and gas.

For ongoing reservoir management, models may help in improved oil recovery by hydraulic fracturing. Highly deviated or horizontal wells can also be represented. Specialized software may be used in the design of hydraulic fracturing, then the improvements in productivity can be included in the field model. Also, future improvement in oil recovery with pressure maintenance by re-injection of produced gas or by water injection into an aquifer can be evaluated. Water flooding resulting in the improved displacement of oil is commonly evaluated using reservoir simulation.

The application of enhanced oil recovery (EOR) processes requires that the field possesses the necessary characteristics to make application successful. Model studies can assist in this evaluation. EOR processes include miscible displacement by natural gas, CO_2, or nitrogen and chemical flooding (polymer, alkaline, surfactant, or a combination of these). Special features in simulation software is needed to represent these processes. In some miscible applications, the "smearing" of the flood front, also called numerical dispersion, may be a problem.

Reservoir simulation is used extensively to identify opportunities to increase oil production in heavy oil deposits. Oil recovery is improved by lowering the oil viscosity by injecting steam or hot water. Typical processes are steam soaks (steam is injected, then oil produced from the same well) and steam flooding (separate steam injectors and oil producers). These processes require simulators with special features to account for heat transfer to the fluids present and the formation, the subsequent property changes and heat losses outside of the formation.

A recent application of reservoir simulation is the modelling of coalbed methane (CBM) production. This application requires a specialized CBM simulator. In addition to the normal fractured (fissured) formation data, CBM simulation requires gas content data values at initial pressure, sorption isotherms, diffusion coefficient, and parameters to estimate the changes in absolute permeability as a function of pore-pressure depletion and gas desorption.

Fundamentals

Traditional finite difference simulators dominate both theoretical and practical work in reservoir simulation. Conventional FD simulation is underpinned by three physical concepts: conservation of mass, isothermal fluid phase behaviour, and the Darcy approximation of fluid flow through porous media. Thermal simulators (most commonly used for heavy crude oil applications) add conservation of energy to this list, allowing temperatures to change within the reservoir.

Numerical techniques and approaches that are common in modern simulators:

- Most modern FD simulation programmes allow for construction of 3-D representations for use in either full-field or single-well models. 2-D approximations are also used in various conceptual models, such as cross-sections and 2-D radial grid models.
- Theoretically, finite difference models permit discretization of the reservoir using both structured and more complex unstructured grids to accurately represent the geometry of the reservoir. Local grid refinements (a finer grid embedded inside of a coarse grid) are also a feature provided by many simulators to more accurately represent the near wellbore multi-phase flow effects. This "refined meshing" near wellbores is extremely important when analyzing issues such as water and gas coning in reservoirs.
- Representation of faults and their transmissibilities are advanced features provided in many simulators. In these models, inter-cell flow transmissibilities must be computed for non-adjacent layers outside of conventional neighbour-to-neighbour connections.
- Natural fracture simulation (known as dual-porosity and dual-permeability) is an advanced feature which model hydrocarbons in tight matrix blocks. Flow occurs from the tight matrix blocks to the more permeable fracture networks that surround the blocks, and to the wells.
- A black oil simulator does not consider changes in composition of the hydrocarbons as the field is produced. The compositional model, is a more complex model, where the PVT properties of oil and gas phases have been fitted to an equation of state (EOS), as a mixture of components. The simulator then uses the fitted EOS equation to dynamically track the movement of both phases and components in field.

The simulation model computes the saturation change of three phases (oil, water and gas) and pressure of each phase in each cell at each time step. As a result of declining pressure as in a reservoir depletion study, gas will be liberated from the oil. If pressures increase as a result of water or gas injection, the gas is re-dissolved into the oil phase.

A simulation project of a developed field, usually requires "history matching" where historical field production and pressures are compared to calculated values. In recent years optimisation tools such as MEPO has helped to accelerate this process, as well as improve the quality of the match obtained. The model's parameters are adjusted until a reasonable match is achieved on a field basis and usually for all wells. Commonly, producing water cuts or water-oil ratios and gas-oil ratios are matched.

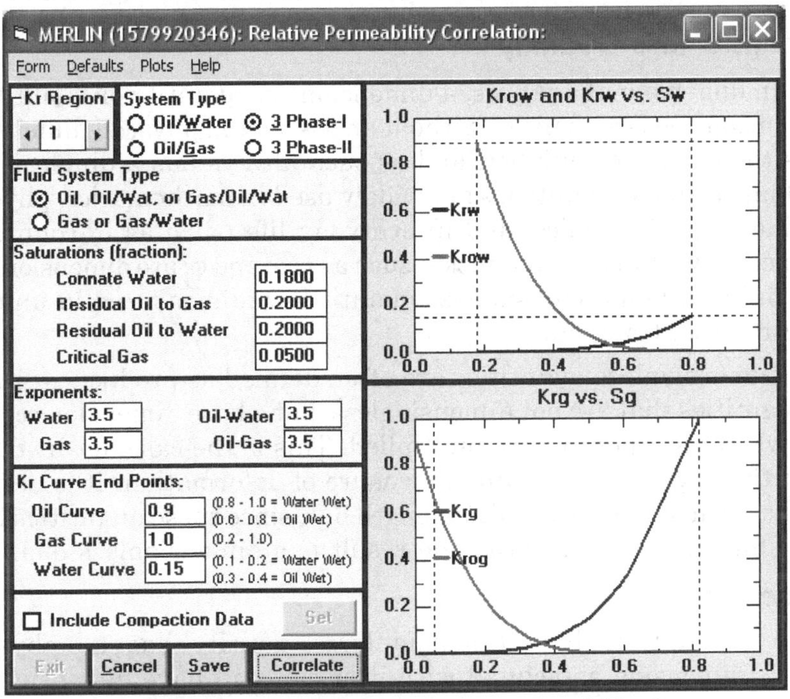

***Figure:** Correlating relative permeability*

Other Engineering Approaches

Without FD models, recovery estimates and oil rates can also be calculated using numerous analytical techniques which include material balance equations (including Havlena-Odeh and Tarner method), fractional flow curve methods (1-D displacement by Buckley-Leverett, Deitz method for inclined structures, coning models), sweep efficiency estimation techniques for water floods and decline curve analysis. These methods were developed and used prior to traditional or "conventional" simulations tools as computationally inexpensive models based on simple homogeneous reservoir description. Analytical methods generally cannot capture all the details of the given reservoir or process, but are typically numerically fast and at times, sufficiently reliable. In modern reservoir engineering, they are generally used as screening or preliminary evaluation tools. Analytical methods are especially suitable for potential assets evaluation when the data are limited and the time is critical, or for broad studies as a pre-screening tool if a large number of processes and / or technologies are to be evaluated. The analytical methods are often developed and promoted in the academia or in-house, however commercial packages also exist.

Dimensionless Quantity

In dimensional analysis, a dimensionless quantity or quantity of dimension one is a quantity without an associated physical dimension. It is thus a "pure" number, and as such always has a dimension of 1. Dimensionless quantities are widely used in mathematics, physics, engineering, economics, and in everyday life (such as in counting). Numerous well-known quantities, such as π, e, and φ, are dimensionless. By contrast, non-dimensionless quantities are measured in units of length, area, time, etc.

Dimensionless quantities are often defined as products or ratios of quantities that are not dimensionless, but whose dimensions cancel out when their powers are multiplied. This is the case, for instance, with the engineering strain, a measure of deformation. It is defined as change in length over initial length but, since these quantities both have dimensions *L* (length), the result is a dimensionless quantity.

Properties

- Even though a dimensionless quantity has no physical dimension associated with it, it can still have dimensionless units. To show the quantity being measured (for example mass fraction or mole fraction), it is sometimes helpful to use the same units in both the numerator and denominator (kg/kg or mol/mol). The quantity may also be given as a ratio of two different units that have the same dimension (for instance, light years over metres). This may be the case when calculating slopes in graphs, or when making unit conversions. Such notation does not indicate the presence of physical dimensions, and is purely a notational convention. Other common dimensionless units are % (= 0.01), ‰ (= 0.001), ppm (= 10^{-6}), ppb (= 10^{-9}), ppt (= 10^{-12}) and angle units (degrees, radians, grad). Units of number such as the dozen and the gross are also dimensionless.

- The ratio of two quantities with the same dimensions is dimensionless, and has the same value regardless of the units used to calculate them. For instance, if body A exerts a force of magnitude F on body B, and B exerts a force of magnitude f on A, then the ratio F/f is always equal to 1, regardless of the actual units used to measure F and f. This is a fundamental property of dimensionless proportions and follows from the assumption that the laws of physics are independent of the system of units used in their expression. In this case, if the ratio F/f was not always equal to 1, but changed if we switched from SI to CGS, that would mean that Newton's Third Law's truth or falsity would depend on the system of units used, which would contradict this fundamental hypothesis. This assumption that the laws of physics are not contingent upon a specific unit system is the basis for the Buckingham π theorem. A statement of this theorem is that any physical law can be expressed as an identity involving only dimensionless combinations (ratios or products) of the variables linked by the law (e. g., pressure and volume are linked by Boyle's Law – they are inversely proportional). If the dimensionless combinations' values changed with the systems of units, then the equation would not be an identity, and Buckingham's theorem would not hold.

Buckingham π Theorem

Another consequence of the Buckingham π theorem of dimensional analysis is that the functional dependence between a certain number (say, n) of variables can be reduced by the number (say, k) of independent dimensions occurring in those variables to give a set of $p = n - k$ independent, dimensionless quantities. For the purposes of the experimenter, different systems that share the same description by dimensionless quantity are equivalent.

Example

The power consumption of a stirrer with a given shape is a function of the density and the viscosity of the fluid to be stirred, the size of the stirrer given by its diameter, and the speed of the stirrer. Therefore, we have $n = 5$ variables representing our example.

Those $n = 5$ variables are built up from $k = 3$ dimensions:

- Length: L (m)
- Time: T (s)
- Mass: M (kg)

According to the π-theorem, the $n = 5$ variables can be reduced by the $k = 3$ dimensions to form $p = n - k = 5 - 3 = 2$ independent dimensionless numbers, which are, in case of the stirrer:

- Reynolds number (a dimensionless number describing the fluid flow regime)
- Power number (describing the stirrer and also involves the density of the fluid)

Standards Efforts

The International Committee for Weights and Measures contemplated defining the unit of 1 as the 'uno', but the idea was dropped.

Examples

- Consider this example: Sarah says, "Out of every 10 apples I gather, 1 is rotten." The rotten-to-gathered ratio is (1 apple) / (10 apples) = 0.1 = 10%, which is a dimensionless quantity.
- Plane angles – An angle is measured as the ratio of the length of a circle's arc subtended by an angle whose vertex is the centre of the circle to some other length. The ratio—i.e., length divided by length—is dimensionless. When using radians as the unit, the length that is compared is the length of the radius of the circle. When using degree as the units, the arc's length is compared to 1/360 of the circumference of the circle.
- In the case of the dimensionless quantity π, being the ratio of a circle's circumference to its diameter, the number would be constant regardless of what unit is used to measure a circle's circumference and diameter (e.g., centimetres, miles, light-years, etc.), as long as the same unit is used for both.

Dimensionless Physical Constants

Certain fundamental physical constants, such as the speed of light in a vacuum, the universal gravitational constant, Planck's constant and Boltzmann's constant can be normalized to 1 if appropriate units for time, length, mass, charge, and temperature are chosen. The resulting system of units is known as the natural units. However, not all physical constants can be normalized in this fashion. For example, the values of the following constants are independent of the system of units and must be determined experimentally:

- $\alpha \approx 1/137.036$, the fine structure constant which is the coupling constant for the electromagnetic interaction;

- β (or μ) $\approx$ 1836, the proton-to-electron mass ratio. This ratio is the rest mass of the proton divided by that of the electron. An analogous ratio can be defined for any elementary particle;
- α_s, the coupling constant for the strong force;
- $\alpha_G \approx 1.75\times10^{-45}$, the gravitational coupling constant.

Skin Effect

"Skin depth" redirects here. For the depth (layers) of biological/ organic skin.

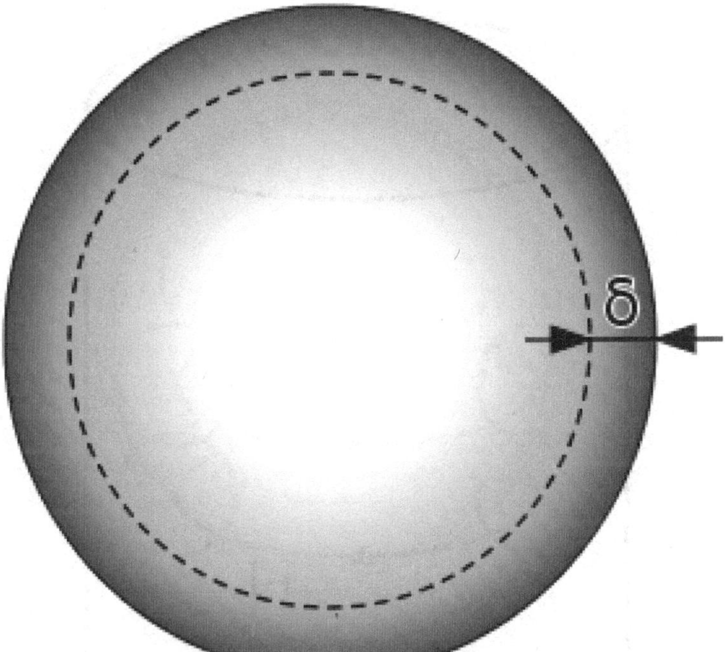

***Figure:** Distribution of current flow in a cylindrical conductor, shown in cross section. For alternating current, most (63%) of the electrical current flows between the surface and the skin depth, δ, which depends on the frequency of the current and the electrical and magnetic properties of the conductor.*

Skin effect is the tendency of an alternating electric current (AC) to become distributed within a conductor such that the current density is largest near the surface of the conductor, and decreases with greater depths in the conductor. The electric current flows mainly at the "skin" of the conductor, between the outer surface and a level called the skin depth. The skin effect causes the effective resistance of the conductor to increase at higher frequencies where the skin depth is smaller, thus reducing the effective cross-section of the conductor. The skin effect is due to opposing eddy currents induced by the changing magnetic field resulting from the alternating current. At 60 Hz in copper, the skin depth is about 8.5 mm. At high frequencies the skin depth becomes much smaller. Increased AC resistance due to the skin effect can be mitigated by using specially woven litz wire. Because the interior of a large conductor carries so little of the current, tubular conductors such as pipe can be used to save weight and cost.

Cause

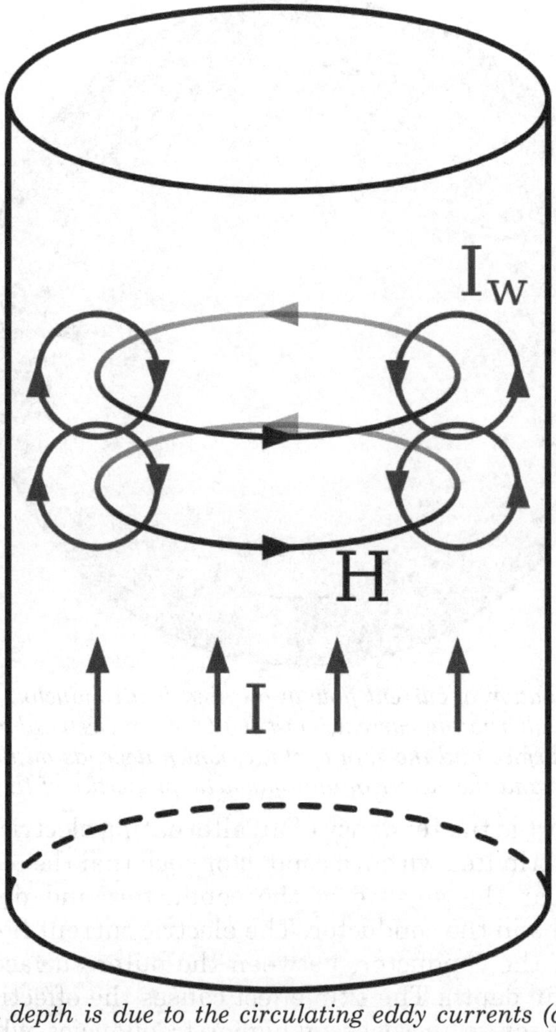

Figure: *Skin depth is due to the circulating eddy currents (arising from a changing H field) cancelling the current flow in the centre of a conductor and reinforcing it in the skin.*

Conductors, typically in the form of wires, may be used to transmit electrical energy or signals using an alternating current flowing through that conductor. The charge carriers constituting that current, usually electrons, are driven by an electric field due to the source of electrical energy. An alternating current in a conductor produces an alternating magnetic field in and around the conductor.

When the intensity of current in a conductor changes, the magnetic field also changes. The change in the magnetic field, in turn, creates an electric field which opposes the change in current intensity. This opposing electric field is called "counter-electromotive force" (back EMF). The back EMF is strongest at the centre of the conductor, and forces the conducting electrons to the outside of the conductor, as shown in the diagram on the right.

An alternating current may also be *induced* in a conductor due to an alternating magnetic field according to the law of induction. An electromagnetic wave impinging on a conductor will therefore generally produce such a current; this explains the reflection of electromagnetic waves from metals.

Regardless of the driving force, the current density is found to be greatest at the conductor's surface, with a reduced magnitude deeper in the conductor. That decline in current density is known as the *skin effect* and the *skin depth* is a measure of the depth at which the current density falls to 1/e of its value near the surface. Over 98% of the current will flow within a layer 4 times the skin depth from the surface. This behaviour is distinct from that of direct current which usually will be distributed evenly over the cross-section of the wire.

The effect was first described in a paper by Horace Lamb in 1883 for the case of spherical conductors, and was generalised to conductors of any shape by Oliver Heaviside in 1885. The skin effect has practical consequences in the analysis and design of radio-frequency and microwave circuits, transmission lines (or waveguides), and antennas. It is also important even at mains frequencies (50 – 60 Hz) in AC electrical power transmission and distribution systems.

Although the term "skin effect" is most often associated with applications involving transmission of electrical currents, the skin depth also describes the exponential decay of the electric and magnetic fields, as well as the density of induced currents, inside a bulk material when a plane wave impinges on it at normal incidence.

Formula

The AC current density J in a conductor decreases exponentially from its value at the surface J_S according to the depth d from the surface, as follows:

$$J = J_S e^{-d/\delta}$$

where δ is called the *skin depth.* The skin depth is thus defined as the depth below the surface of the conductor at which the current density has fallen to 1/e (about 0.37) of J_S. In normal cases it is well approximated as:

$$\delta = \sqrt{\frac{2\rho}{\omega\mu}}.$$

where

ρ = resistivity of the conductor

ω = angular frequency of current = $2\pi \times$ frequency

μ = absolute magnetic permeability of the conductor

A more general expression for skin depth which is more exact in the case of poor conductors (non-metals) at high frequencies is:

$$\delta = \left(\frac{1}{\omega}\right)\left\{\left(\frac{\mu\epsilon}{2}\right)\left[\left(1+\left(\frac{1}{\rho\omega\epsilon}\right)^2\right)^{1/2} - 1\right]\right\}^{-1/2}$$

where ϵ is the electric permittivity of the material. Note that in the usual form for the skin effect, above, the effect of cancels out. This formula is valid away from strong atomic or molecular resonances (where would have a large imaginary part) and at frequencies which are much below both the material's plasma frequency (dependent on the density of free electrons in the material) and the reciprocal of the mean time between collisions involving the conduction electrons. In good conductors such as metals all of those conditions are ensured at least up to microwave frequencies, justifying this formula's validity.

This formula can be rearranged as follows to reveal departures from the normal approximation:

$$\delta = \sqrt{\frac{2\rho}{\omega\mu}}\sqrt{\sqrt{1+(\rho\omega\epsilon)^2}+\rho\omega\epsilon}$$

At frequencies much below $1/\rho\epsilon$ the quantity inside the radical is close to unity and the standard formula applies. For instance, in the case of copper this would be true for frequencies much below 10^{18} Hz.

However in very poor conductors at sufficiently high frequencies, the factor on the right increases. At frequencies much higher than $1/\rho\epsilon$ it can be shown that the skin depth, rather than continuing to decrease, approaches an asymptotic value:

$$\delta \approx 2\rho\sqrt{\frac{\epsilon}{\mu}} \qquad (\omega \gg 1/\rho\epsilon)$$

This departure from the usual formula only applies for materials of rather low conductivity and at frequencies where the vacuum wavelength is not much much larger than the skin depth itself. For instance, bulk silicon (undoped) is a poor conductor and has a skin depth of about 40 metres at 100 kHz (λ =3000m). However as the frequency is increased well into the megahertz range, its skin depth never falls below the asymptotic value of 11 metres. The conclusion is that in poor solid conductors such as undoped silicon, the skin effect doesn't need to be taken into account in most practical situations: any current is equally distributed throughout the material's cross-section regardless of its frequency.

Resistance

The effective resistance due to a current confined near the surface of a large conductor (much thicker than δ) can be solved as if the current flowed uniformly through a layer of thickness δ based on the DC resistivity of that material. We can therefore assume a cross-sectional area approximately equal to δ times the conductor's circumference. Thus a long cylindrical conductor such as a wire, having a diameter D large compared to δ, has a resistance *approximately* that of a hollow tube with wall thickness δ carrying direct current. Using a material of resistivity we then find the AC resistance of a wire of length L to be:

$$R \approx \frac{L\rho}{\pi(D-\delta)\delta} \approx \frac{L\rho}{\pi D\delta}$$

The final approximation above assumes $D \gg \delta$.

A convenient formula (attributed to F.E. Terman) for the diameter D_W of a wire of circular cross-section whose resistance will increase by 10% at frequency f is:

$$D_W = \frac{200\,\text{mm}}{\sqrt{f/\text{Hz}}}$$

The increase in AC resistance described above is accurate only for an isolated wire. For a wire close to other wires, e.g. in a cable or a coil, the ac resistance is also affected by proximity effect, which often causes a much more severe increase in ac resistance.

Material Effect on Skin Depth

In a good conductor, skin depth varies as the inverse square root of the conductivity. This means that better conductors have a reduced skin depth. The overall resistance of the better conductor remains lower even with the reduced skin depth. However the better conductor will show a higher ratio between its AC and DC resistance, when compared with a conductor of higher resistivity. For example, at 60 Hz, a 2000 MCM (1000 square millimetre) copper conductor has 23% more resistance than it does at DC. The same size conductor in aluminium has only 10% more resistance with 60 Hz AC than it does with DC.

Skin depth also varies as the inverse square root of the permeability of the conductor. In the case of iron, its conductivity is about 1/7 that of copper. However being ferromagnetic its permeability is about 10,000 times greater. This reduces the skin depth for iron to about 1/38 that of copper, about 220 micrometres at 60 Hz. Iron wire is thus useless for A.C. power lines. The skin effect also reduces the effective thickness of laminations in power transformers, increasing their losses.

Iron rods work well for direct-current (DC) welding but it is impossible to use them at frequencies much higher than 60 Hz. At a few kilohertz, the welding rod will glow red hot as current flows through the greatly increased A.C. resistance resulting from the skin effect, with relatively little power remaining for the arc itself. Only non-magnetic rods can be used for high-frequency welding.

Mitigation

A type of cable called litz wire (from the German *Litzendraht*, braided wire) is used to mitigate the skin effect for frequencies of a few kilohertz to about one megahertz. It consists of a number of insulated wire strands woven together in a carefully designed pattern, so that the overall magnetic field acts equally on all the wires and causes the total current to be distributed equally among them. With the skin effect having little effect on each of the thin strands, the bundle does not suffer the same increase in AC resistance that a solid conductor of the same cross-sectional area would due to the skin effect.

Litz wire is often used in the windings of high-frequency transformers to increase their efficiency by mitigating both skin effect and proximity effect. Large power transformers are wound with stranded conductors of similar construction to litz wire, but employing a larger cross-section corresponding to the larger skin depth at mains frequencies. Conductive threads composed of carbon nanotubes have been demonstrated as conductors for antennas from medium wave to microwave frequencies. Unlike standard antenna conductors, the nanotubes are much smaller than the skin depth, allowing full utilization of the thread's cross-section resulting in an extremely light antenna.

High-voltage, high-current overhead power lines often use aluminium cable with a steel reinforcing core; the higher resistance of the steel core is of no consequence since it is located far below the skin depth where essentially no AC current flows.

In other applications where high currents (up to many thousands of amperes) flow, where practicable solid conductors are usually replaced by tubes, completely dispensing with the inner portion of the conductor where little current flows. This hardly affects the AC resistance but considerably reduces the weight of the conductor. The high strength but low weight of tubes substantially increases span capability. Hence tubular conductors are typical to electric power switchyards where the distance between supporting insulators may be several metres. Long spans generally exhibit physical sag but this does not affect electrical performance. However to avoid losses, the conductivity of the tube material must be high. To this end, care is needed in alloy selection, because the presence of a small portion of impurities in the base metal or alloy will reduce conductivity dramatically when compared to the same metal in pure form. e.g. copper tube produced for plumbing purposes includes additives that reduce electrical conductivity to a low level.

Tubular conductors also present challenges in jointing (usually welded). At terminations it is usually necessary to weld a flat lug or tag to the tube to enable connection to a flat busbar or aerial conductor at some point in the circuit.

In high current situations where conductors (round or flat busbar) may be between 5 and 50 mm thick the skin effect also occurs at sharp bends where the metal is compressed inside the bend and stretched outside the bend. The shorter path at the inner surface results in a lower resistance, which causes most of the current to be concentrated close to the inner bend surface. This will cause an increase in temperature rise in that region compared with the straight (unbent) area of the same conductor. A similar skin effect occurs at the corners of rectangular conductors (viewed in cross-section), where the magnetic field is more concentrated at the corners than in the sides. This results in superior performance (i.e. higher current with lower temperature rise) from wide thin conductors - e.g. "ribbon" conductors, where the effects from corners is effectively eliminated.

It follows that a transformer having a round core will be more efficient than an equivalent VA rated transformer having a square or rectangular core of the same material.

Solid or tubular conductors may also be silver-plated to take advantage of silver's higher conductivity. This technique is particularly used at VHF to microwave frequencies where the small skin depth requires only a very thin layer of silver, making the improvement in conductivity very cost effective. Silver or gold plating is similarly used on the surface of waveguides used for transmission of microwaves. This reduces attenuation of the propagating wave due to resistive losses affecting the accompanying eddy currents; the skin effect confines such eddy currents to a very thin surface layer of the waveguide structure. The skin effect itself isn't actually combated in these cases, but the distribution of currents near the conductor's surface makes the use of precious metals (having a lower resistivity) practical. (Note commercial high-conductivity copper has an electrical conductivity superior to silver but is more susceptible to corrosion)

NOTE: The heat dependent manufacturing process for wire results in oxidation of the surface in the finished product. Therefore the conductivity of the skin is significantly less than the underlying unoxidised base metal. It is to be expected that unless the oxidised surface material is removed, some loss of performance from the theoretical model will occur.

Examples

Mn-Zn - magnetically soft ferrite

Al - metallic alumium

Cu - metallic copper

steel 410 - magnetic stainless steel

Fe-Si - grain-oriented electrical steel

Fe-Ni - high-permeability permalloy (80%Ni-20%Fe)

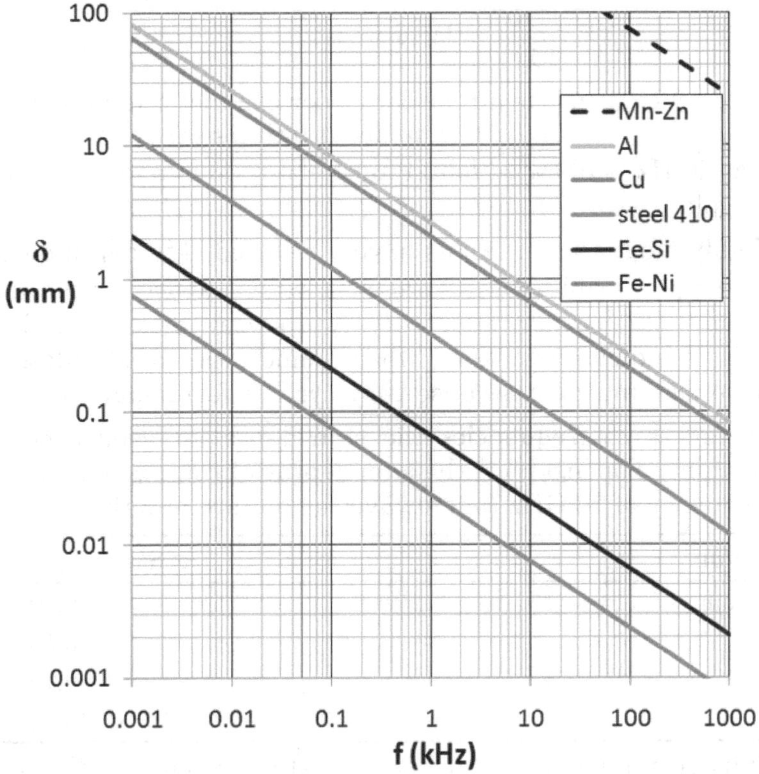

Figure: *Skin depth vs. frequency for some materials, red vertical line denotes 50 Hz frequency*

We can derive a practical formula for skin depth as follows:

$$\delta = \sqrt{\frac{2\rho}{(2\pi f)(\mu_0 \mu_r)}} \approx 503\sqrt{\frac{\rho}{\mu_r f}}$$

where

δ = the skin depth in metres

μ_r = the relative permeability of the medium

ρ = the resistivity of the medium in $\Omega \cdot$m, also equal to the reciprocal of its conductivity:

$\rho = 1/\sigma$ (for Copper, $\rho = 1.68\times10^{-8}\ \Omega\cdot$m)

f = the frequency of the current in Hz

Gold is a good conductor with a resistivity of $2.44\times10^{-8}\ \Omega\cdot$m and is essentially nonmagnetic: $\mu_r = 1$, so its skin depth at a frequency of 50 Hz is given by

$$\delta = 503\sqrt{\frac{2.44 \cdot 10^{-8}}{1 \cdot 50}} = 11.1\text{mm}$$

Lead, in contrast, is a relatively poor conductor (among metals) with a resistivity of 2.2×10^{-7} Ω·m, about 9 times that of gold. Its skin depth at 50 Hz is likewise found to be about 33 mm, or $\sqrt{9} = 3$ times that of gold.

Highly magnetic materials have a reduced skin depth owing to their large permeability μ_r as was pointed out above for the case of iron, despite its poorer conductivity. A practical consequence is seen by users of induction cookers, where some types of stainless steel cookware are unusable because they are not ferromagnetic.

At very high frequencies the skin depth for good conductors becomes tiny. For instance, the skin depths of some common metals at a frequency of 10 GHz (microwave region) are less than a micron:

Conductor	***Skin depth (mm)***
Aluminium	0.80
Copper	0.65
Gold	0.79
Silver	0.64

Thus at microwave frequencies, most of the current flows in an extremely thin region near the surface. Ohmic losses of waveguides at microwave frequencies are therefore only dependent on the surface coating of the material. A layer of silver 3 μm thick evaporated on a piece of glass is thus an excellent conductor at such frequencies.

In copper, the skin depth can be seen to fall according to the square root of frequency:

Frequency	***Skin depth (mm)***
60 Hz	8470
10 kHz	660
100 kHz	210
1 MHz	66
10 MHz	21
100 MHz	6.6

In *Engineering Electromagnetics*, Hayt points out that in a power station a busbar for alternating current at 60 Hz with a radius larger than one-third of an inch (8 mm) is a waste of copper, and in practice bus bars for heavy AC current are rarely more than half an inch (12 mm) thick except for mechanical reasons.

Skin Effect Reduction of the Self Inductance of a Conductor

Refer to the diagram below showing the inner and outer conductors of a coaxial cable. Since the skin effect causes a current at high frequencies to flow mainly at the surface of a conductor, it can be seen that this will reduce the magnetic field *inside* the wire, that is, beneath the depth at which the bulk of the current flows. It can be shown that this will have a minor effect on the self-inductance of the wire itself.

Note that the inductance considered in this context refers to a bare conductor, not the inductance of a coil used as a circuit element. The inductance of a coil is dominated by the mutual inductance between the turns of the coil which increases its inductance according to the square of the number of turns. However when only a single wire is involved, then in addition to the "external inductance" involving magnetic fields outside of the wire (due to the total current in the wire), there is also a much smaller component of "internal inductance" due to the magnetic field inside the wire itself, the green region. In a single wire the internal inductance becomes of little significance when the wire is much much longer than its diameter. The presence of a second conductor in the case of a transmission line requires a different treatment as is discussed below.

Due to the skin effect, at high frequencies the internal inductance of a wire vanishes, as can be seen in the case of a telephone twisted pair, below. In normal cases the effect of internal inductance is ignored in the design of coils or calculating the properties of microstrips.

Inductance Per Length in a Coaxial Cable

Let the dimensions a, b, and c be the inner conductor radius, the shield (outer conductor) inside radius and the shield outer radius respectively..

For a given current, the total energy stored in the magnetic fields must be the same as the calculated electrical energy attributed to that current flowing through the inductance of the coax; that energy is proportional to the cable's measured inductance.

The magnetic field inside a coaxial cable can be divided into three regions, each of which will therefore contribute to the electrical inductance seen by a length of cable.

The inductance L_{cen} is associated with the magnetic field in the region with radius $r < a$, the region inside the centre conductor.

The inductance L_{ext} is associated with the magnetic field in the region $a < r < b$, the region between the two conductors (containing a dielectric, possibly air).

The inductance L_{shd} is associated with the magnetic field in the region $b < r < c$, the region inside the shield conductor.

The net electrical inductance is due to all three contributions:

$$L_{total} = L_{cen} + L_{shd} + L_{ext}$$

L_{ext} is not changed by the skin effect and is given by the frequently cited formula for inductance *L* per length *D* of a coaxial cable:

$$L / D = \frac{\mu_0}{2\pi} \ln\left(\frac{b}{a}\right)$$

At low frequencies, all three inductances are fully present so that $L_{DC} = L_{cen} + L_{shd} + L_{ext}$.

At high frequencies, only the dielectric region has magnetic flux, so that $L_\infty = L_{ext}$.

Most discussions of coaxial transmission lines assume they will be used for radio frequencies, so equations are supplied corresponding only to the latter case.

As the skin effect increases, the currents are concentrated near the outside of the inner conductor (*r=a*) and the inside of the shield (*r=b*). Since there is essentially no current deeper in the inner conductor, there is no magnetic field beneath the surface of the inner conductor.

Since the current in the inner conductor is balanced by the opposite current flowing on the inside of the outer conductor, there is no remaining magnetic field in the outer conductor itself where $b < r < c$. L_{ext} Only contributes to the electrical inductance at these higher frequencies.

Although the geometry is different, a twisted pair used in telephone lines is similarly affected: at higher frequencies the inductance decreases by more than 20% as can be seen in the following table.

Characteristics of Telephone Cable as a Function of Frequency

Representative parameter data for 24 gauge PIC telephone cable at 21 °C (70 °F).

Frequency (Hz)	*R (Ω/km)*	*L (mH/km)*	*G (μS/km)*	*C (nF/km)*
1	172.24	0.6129	0.000	51.57
1k	172.28	0.6125	0.072	51.57
10k	172.70	0.6099	0.531	51.57
100k	191.63	0.5807	3.327	51.57
1M	463.59	0.5062	29.111	51.57
2M	643.14	0.4862	53.205	51.57
5M	999.41	0.4675	118.074	51.57

More extensive tables and tables for other gauges, temperatures and types are available in Reeve. Chen gives the same data in a parametreized form that he states is usable up to 50 MHz.

Chen gives an equation of this form for telephone twisted pair:

$$L(f) = \frac{l_0 + l_\infty (\frac{f}{f_m})^b}{1 + (\frac{f}{f_m})^b}$$

Flow Coefficient

The flow coefficient of a device is a relative measure of its efficiency at allowing fluid flow. It describes the relationship between the pressure drop across an orifice, valve or other assembly and the corresponding flow rate.

Mathematically the flow coefficient can be expressed as:

$$C_v = F\sqrt{\frac{SG}{\Delta P}}$$

where:

C_v = Flow coefficient or flow capacity rating of valve.

F = Rate of flow (US gallons per minute).

SG = Specific gravity of fluid (Water = 1).

ΔP = Pressure drop across valve (psi).

In more practical terms, the *flow coefficient* C_v is the volume (in US gallons) of water at 60°F that will flow per minute through a valve with a pressure drop of 1 psi across the valve.

The use of the flow coefficient offers a standard method of comparing valve capacities and sizing valves for specific applications that is widely accepted by industry. The general definition of the flow coefficient can be expanded into equations modelling the flow of liquids, gases and steam as follows:

Coefficient of discharge is the ratio of actual flow rate to theoretical discharge. For gas flow in a pneumatic system the C_v for the same assembly can be used with a more complex equation. Absolute pressures (psia) must be used for gas rather than simply differential pressure.

For air flow at room temperature, when the outlet pressure is less than 1/2 the absolute inlet pressure, the flow becomes quite simple (although it reaches sonic velocity internally). With C_v = 1.0 and 200 psia inlet pressure the flow is 100 standard cubic feet per minute (scfm). The flow is proportional to the absolute inlet pressure so that the flow in scfm would equal the C_v flow coefficient if the inlet pressure were reduced to 2 psia and the outlet were connected to a vacuum with less than 1 psi absolute pressure (1.0 scfm when C_v = 1.0, 2 psia input).

Semi-log Plot

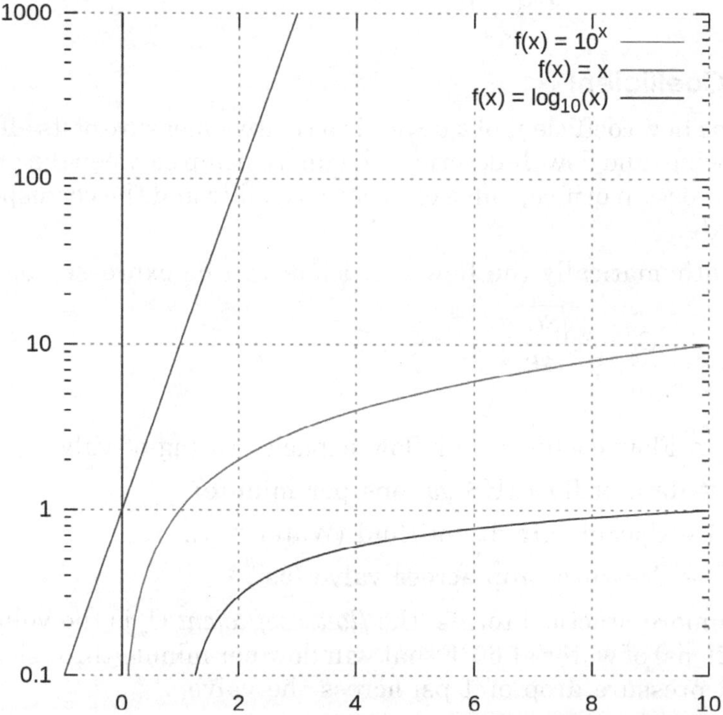

Figure: *The log-lin type of a semi-log graph, defined by a logarithmic scale on the y-axis, and a linear scale on the x-axis. Plotted lines are: y = 10^x (red), y = x (green), y = log(x) (blue).*

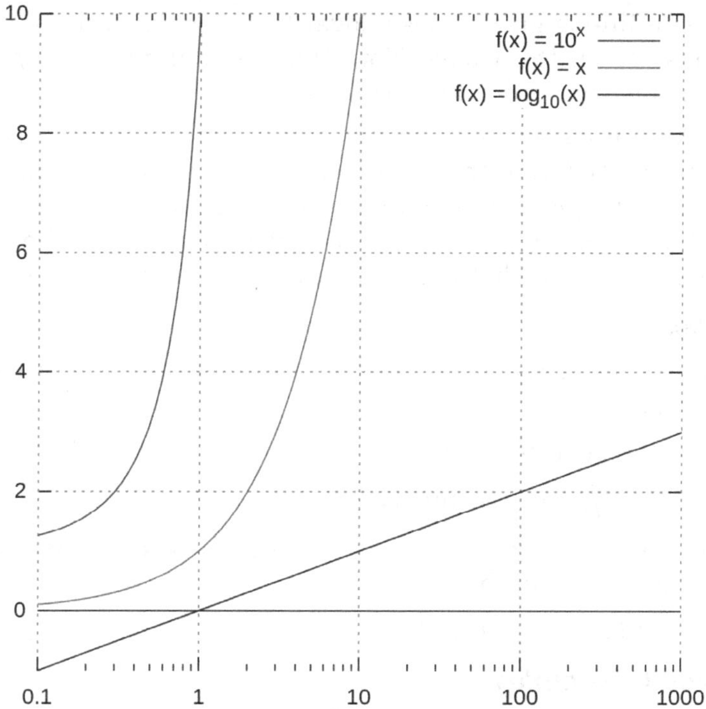

Figure: *The lin-log type of a semi-log graph, defined by a logarithmic scale on the x axis, and a linear scale on the y axis. Plotted lines are:* y = 10^x *(red),* y = x *(green),* y = *log*(x) *(blue).*

In science and engineering, a semi-log graph or semi-log plot is a way of visualizing data that are related according to an exponential relationship. One axis is plotted on a logarithmic scale. This kind of plot is useful when one of the variables being plotted covers a large range of values and the other has only a restricted range – the advantage being that it can bring out features in the data that would not easily be seen if both variables had been plotted linearly.

All equations of the form $y = \lambda a^{\gamma x}$ form straight lines when plotted semi-logarithmically, since taking logs of both sides gives

$$\log_a y = \gamma x + \log_a \lambda.$$

This can easily be seen as a line in slope-intercept form with γ as the slope and $\log_a \lambda$ as the vertical intercept. To facilitate use with logarithmic tables, one usually takes logs to base 10 or e, or sometimes base 2:

$$\log(y) = (\gamma \log(a))x + \log(\lambda).$$

The term log-lin is used to describe a semi-log plot with a logarithmic scale on the *y*-axis, and a linear scale on the *x*-axis.

Likewise, a lin-log graph uses a logarithmic scale on the x-axis, and a linear scale on the y-axis. Note that the naming is *output-input* (y-x), the opposite order from (x, y).

On a semi-log graph the spacing of the scale on the y-axis (or x-axis) is proportional to the logarithm of the number, not the number itself. It is equivalent to converting the y values (or x values) to their log, and plotting the data on lin-lin scales. A log-log graph uses the logarithmic scale for both axes, and hence is not a semi-log graph.

Equations

The equation for a line with an ordinate axis logarithmically scaled would be:

$$\log_{10}(F(x)) = mx + b$$

$$F(x) = 10^{mx+b} = (10^{mx})(10^{b}).$$

The equation of a line on a plot where the abscissa axis is scaled logarithmically would be

$$F(x) = m\log_{10}(x) + b.$$

Real-world Examples

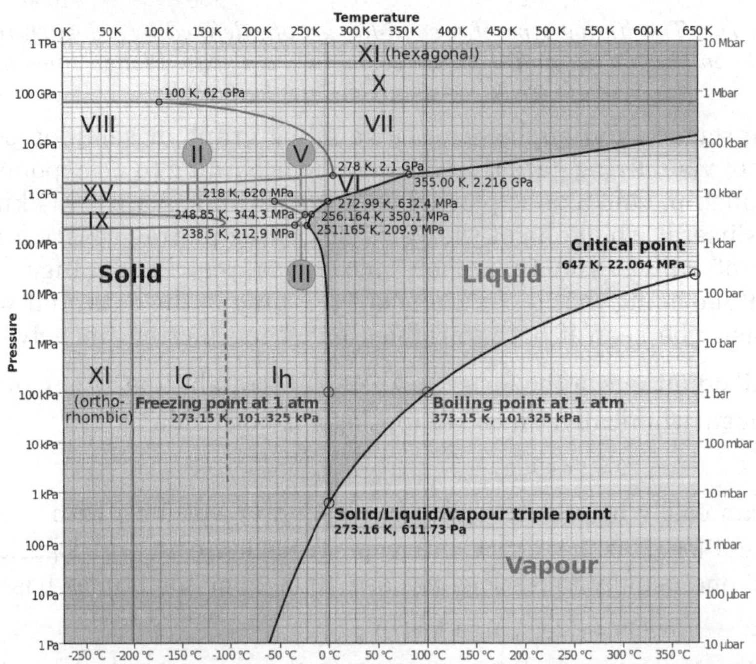

Figure: *Log-lin pressure–temperature phase diagram of water. The Roman numerals indicate various ice phases.*

Phase Diagram of Water: In physics and chemistry, a plot of logarithm of pressure against temperature can be used to illustrate the various phases of a substance.

2009 "Swine flu" Progression

While ten is the most common base, there are times when other bases are more appropriate, as in this example:

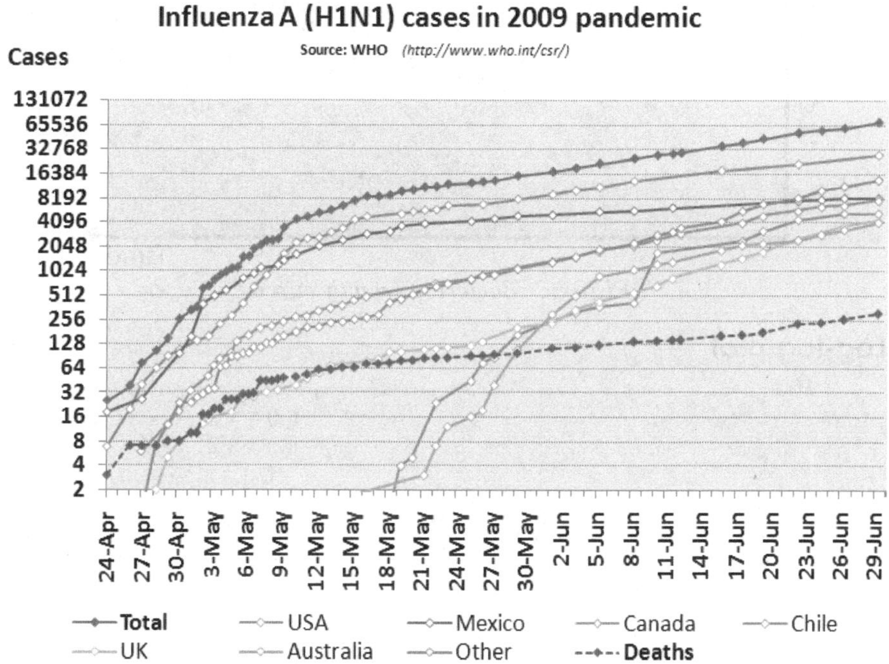

Figure: *A semi-logarithmic plot of cases and deaths in the 2009 outbreak of influenza A (H1N1). Notice that while the horizontal (time) axis is linear, with the dates evenly spaced, the vertical (cases) axis is logarithmic, with the evenly spaced divisions being labelled with successive powers of two.*

Microbial Growth

In biology and biological engineering, the change in numbers of microbes due to asexual reproduction and nutrient exhaustion is commonly illustrated by a semi-log plot.

Time is usually the independent axis, with the logarithm of the number or mass of bacteria or other microbe as the dependent variable. This forms a plot with four distinct phases, as shown below.

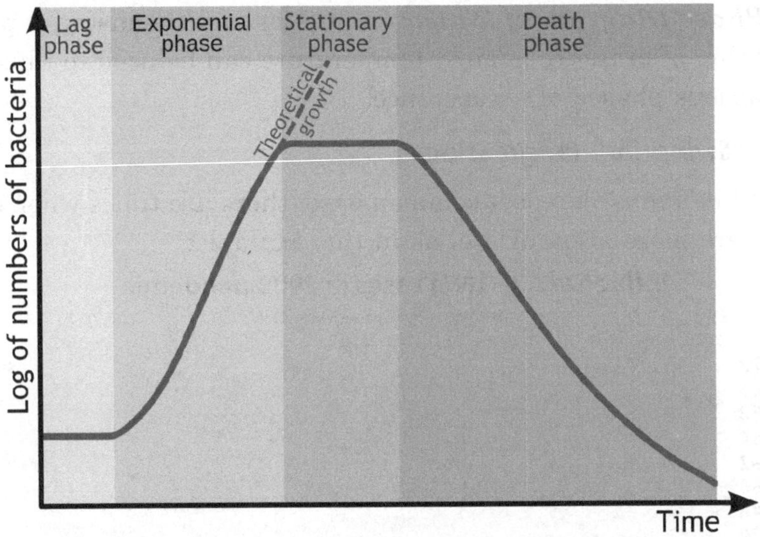

Figure: *Bacterial growth curve*

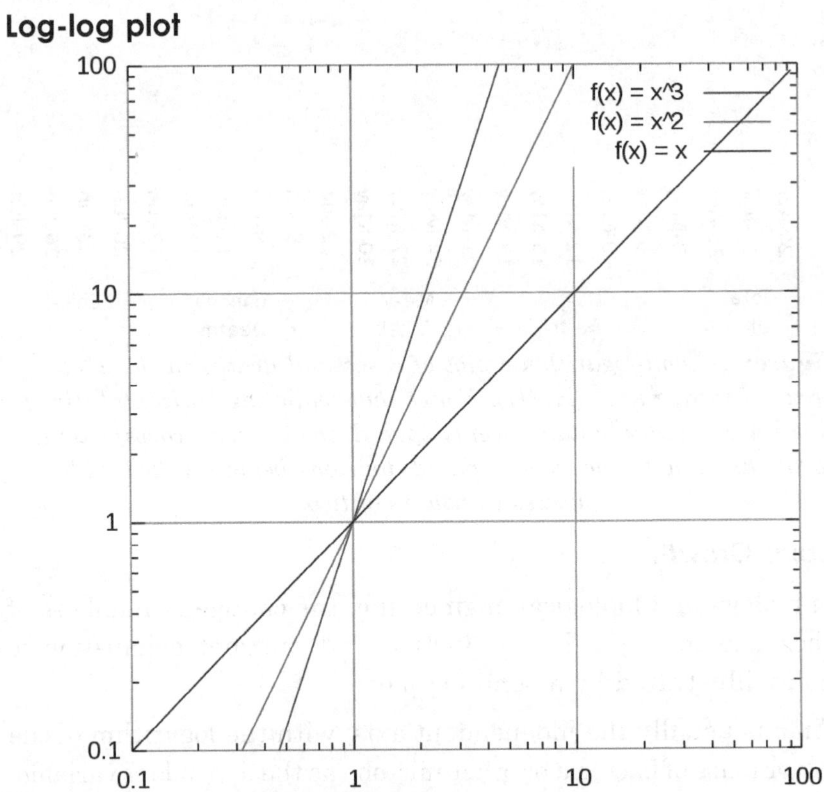

Figure: *A log-log plot of* y = x *(blue),* $y = x^2$ *(green), and* $y = x^3$ *(red).*

Note the logarithmic scale markings on each of the axes, and that the log x and log y axes (where the logarithms are 0) are where x and y themselves are 1.

In science and engineering, a log-log graph or log-log plot is a two-dimensional graph of numerical data that uses logarithmic scales on both the horizontal and vertical axes. Monomials – relationships of the form $y = ax^k$ – appear as straight lines in a log-log graph, with the power and constant term corresponding to slope and intercept of the line, and thus these graphs are very useful for recognizing these relationships and estimating parameters. Any base can be used for the logarithm, though most common are 10, e, and 2.

Relation with Monomials

Given a monomial equation $y = ax^k$, taking the logarithm of the equation (with any base) yields:

$$(\log y) = k(\log x) + \log a.$$

Setting $X = \log x$ and $Y = \log y$, which corresponds to using a log-log graph, yields the equation:

$$Y = mX + b$$

where $m = k$ is the slope of the line (gradient) and $b = \log a$ is the intercept on the (log y)-axis, meaning where log x = 0, so, reversing the logs, a is the y value corresponding to x = 1.

Equations

The equation for a line on a log-log scale would be:

$$\log_{10}(F(x)) = m\log_{10}(x) + b,$$
$$F(x) = (x^m)(10^b),$$

where m is the slope and b is the intercept point on the log plot.

Slope of a Log-log Plot

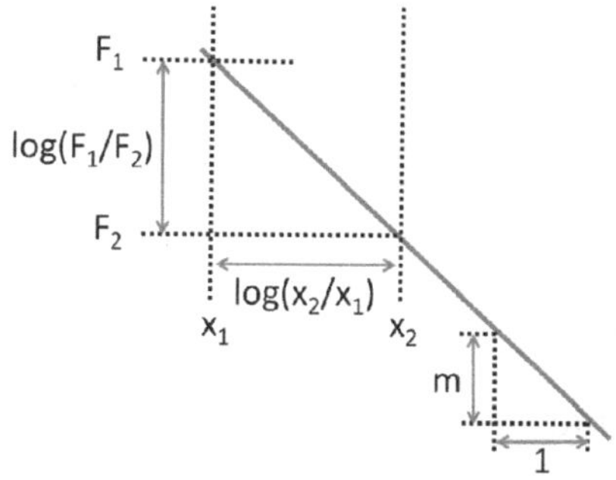

Figure: *Finding the slope of a log-log plot using ratios*

To find the slope of the plot, two points are selected on the x-axis, say x_1 and x_2. Using the above equation:

$$\log[F(x_1)] = m\log(x_1) + b,$$

and

$$\log[F(x_2)] = m\log(x_2) + b.$$

The slope m is found taking the difference:

$$m = \frac{\log(F_2) - \log(F_1)}{\log(x_2) - \log(x_1)} = \frac{\log(F_2 / F_1)}{\log(x_2 / x_1)},$$

where F_1 is shorthand for $F(x_1)$ and the same for F_2. The figure illustrates the formula. Notice that the slope in the example of the figure is *negative*. The formula also provides a negative slope, as can be seen from the following property of the logarithm:

$$\log(x_1 / x_2) = -\log(x_2 / x_1).$$

Finding the Function from the Log-log Plot

The above procedure now is reversed to find the form of the function $F(x)$ using its (assumed) known log-log plot. To find the function F, pick some *fixed point* (x_0, F_0), where F_0 is shorthand for $F(x_0)$, somewhere on the straight line in the above graph, and further some other *arbitrary point* (x_1, F_1) on the same graph. Then from the slope formula above:

$$m = \frac{\log(F_1 / F_0)}{\log(x_1 / x_0)}$$

which leads to

$$\log(F_1 / F_0) = m\log(x_1 / x_0) = \log[(x_1 / x_0)^m].$$

Notice that $10^{\log_{10}(F_1)} = F_1$. Therefore, the logs can be inverted to find:

$$\frac{F_1}{F_0} = \left(\frac{x_1}{x_0}\right)^m$$

or

$$F_1 = \frac{F_0}{x_0^m} x^m,$$

which means that

$$F(x) = \text{constant} \cdot x^m.$$

In other words, F is proportional to x to the power of the slope of the straight line of its log-log graph. Specifically, a straight line on a log-log plot containing points (F_0, x_0) and (F_1, x_1) will have the function:

$$F(x) = F_0 \left(\frac{x}{x_0} \right)^{\frac{\log(F_1/F_0)}{\log(x_1/x_0)}},$$

Of course, the inverse is true too: any function of the form

$$F(x) = \text{constant} \cdot x^m$$

will have a straight line as its log-log graph representation, where the slope of the line is m.

Applications

These graphs are useful when the parameters a and b need to be estimated from numerical data. Specifications such as this are used frequently in economics. One example is the estimation of money demand functions based on inventory theory, in which it can be assumed that money demand at time t is given by

$$M_t = AR_t^b Y_t^c U_t,$$

where M is the real quantity of money held by the public, R is the rate of return on an alternative, higher yielding asset in excess of that on money, Y is the public's real income, U is an error term assumed to be lognormally distributed, A is a scale parameter to be estimated, and b and c are elasticity parameters to be estimated. Taking logs yields

$$m_t = a + br_t + cy_t + u_t,$$

where m=log M, a = log A, r = log R, y = log Y, and u = log U with u being normally distributed. This equation can be estimated using ordinary least squares.

Another economic example is the estimation of a firm's Cobb-Douglas production function, which is the right side of the equation

$$Q_t = AN_t^{\alpha} K_t^{\beta} U_t,$$

in which Q is the quantity of output that can be produced per month, N is the number of hours of labour employed in production per month, K is the number of hours of physical capital utilized per month, U is an error term assumed to be lognormally distributed, and A, α, and $\beta$are parameters to be estimated. Taking logs gives the linear regression equation

$$q_t = a + \alpha n_t + \beta k_t + u_t$$

where q = log Q, a=log A, n=log N, k=log K, and u=log U.

Log-log regression can also be used to estimate the fractal dimension of a naturally occurring fractal.

However, going in the other direction – observing that data appears as an approximate line on a log-log scale and concluding that the data follows a power law – is invalid. In fact, many other functional forms appear approximately linear on the log-log scale, and simply evaluating the goodness of fit of a linear regression on logged data using the coefficient of determination (R^2) may be invalid, as the assumptions of the linear regression model, such as Gaussian error, may not be satisfied; in addition, tests of fit of the log-log form may exhibit low statistical power, as these tests may have low likelihood of rejecting power laws in the presence of other true functional forms.

While simple log-log plots may be instructive in detecting possible power laws, and have been used dating back to Pareto in the 1890s, validation as a power laws requires more sophisticated statistics.

These graphs are also extremely useful when data are gathered by varying the control variable along an exponential fashion, in which case the control variable x is more naturally represented on a log scale, so that the data points are evenly spaced, rather than compressed at the low end.

The output variable y can either be represented linearly, yielding a lin-log graph (log x, y), or its logarithm can also be taken, yielding the log-log graph (log x, log y).

Hydraulic Fracturing

Hydraulic fracturing is the fracturing of rock by a pressurized liquid. Some hydraulic fractures form naturally—certain veins or dikes are examples. Induced hydraulic fracturing or hydrofracturing, commonly known as fracking, is a technique in which typically water is mixed with sand and chemicals, and the mixture is injected at high pressure into a wellbore to create small fractures (typically less than 1mm), along which fluids such as gas, petroleum and brine water may migrate to the well.

The radial distance of influence of the process from the well bore is typically 150 yards. Hydraulic pressure is removed from the well, then small grains of proppant (sand or aluminium oxide) hold these fractures open once the rock achieves equilibrium. The technique is very common in wells for shale gas, tight gas, tight oil, and coal seam gas and hard rock wells. This well stimulation is only conducted once in the life of the well and greatly enhances fluid removal and well productivity. A different technique where only acid is injected is referred to as acidizing.

The first experimental use of hydraulic fracturing was in 1947, and the first commercially successful applications were in 1949. As of 2010, it was estimated that 60% of all new oil and gas wells worldwide were being hydraulically fractured.

As of 2012, 2.5 million hydraulic fracturing jobs have been performed on oil and gas wells worldwide, more than one million of them in the United States.

Figure: *Halliburton Frack Job in the Bakken*

Proponents of hydraulic fracturing point to the economic benefits from the vast amounts of formerly inaccessible hydrocarbons the process can extract. Opponents point to potential environmental impacts, including contamination of ground water, depletion of fresh water, risks to air quality, the migration of gases and hydraulic fracturing chemicals to the surface, surface contamination from spills and flow-back, and the health effects of these. For these reasons hydraulic fracturing has come under international scrutiny, with some countries suspending or banning it.

However, some of those countries, including most notably the United Kingdom, have recently lifted their bans, choosing to focus on regulations instead of outright prohibition. Documented groundwater contamination has occurred from seepage of the stored water from the hydraulic fracturing from unlined surface ponds.

The 2013 draft EU-Canada trade treaty includes language outlawing any "breach of legitimate expectations of investors" which may occur if revoking drilling licences of Canada-registered companies in the territory of the European Union after the treaty comes into force.

Mechanics

Fracturing in rocks at depth tends to be suppressed by the confining pressure, due to the load caused by the overlying rock strata and the cementation of the formation. This is particularly so in the case of "tensile" (Mode 1) fractures, which require the walls of the fracture to move apart, working against this confining pressure.

Hydraulic fracturing occurs when the effective stress is overcome sufficiently by an increase in the pressure of fluids within the rock, such that the minimum principal stress becomes tensile and exceeds the tensile strength of the material.

Fractures formed in this way will in the main be oriented in the plane perpendicular to the minimum principal stress and for this reason induced hydraulic fractures in well bores are sometimes used to determine the orientation of stresses.

In natural examples, such as dikes or vein-filled fractures, the orientations can be used to infer past states of stress.

Veins

Most vein systems are a result of repeated hydraulic fracturing during periods of relatively high pore fluid pressure. This is particularly noticeable in the case of "crack-seal" veins, where the vein material can be seen to have been added in a series of discrete fracturing events, with extra vein material deposited on each occasion.

One mechanism to demonstrate such examples of long-lasting repeated fracturing is the effect of seismic activity, in which the stress levels rise and fall episodically and large volumes of connate water may be expelled from fluid-filled fractures during earthquakes.

This process is referred to as "seismic pumping".

Dikes

Low-level minor intrusions such as dikes propagate through the crust in the form of fluid-filled cracks, although in this case the fluid is magma. In sedimentary rocks with a significant water content the fluid at the propagating fracture tip will be steam.

Non-hydraulic Fracturing

Fracturing as a method to stimulate shallow, hard rock oil wells dates back to the 1860s. It was applied by oil producers in the US states of Pennsylvania, New York, Kentucky, and West Virginia by using liquid and later also solidified nitroglycerin. Later, the same method was applied to water and gas wells. The idea to use acid as a nonexplosive fluid for well stimulation was introduced in the 1930s. Due to acid etching, fractures would not close completely and therefore productivity was enhanced.

Hydraulic Fracturing in oil and Gas Wells

The relationship between well performance and treatment pressures was studied by Floyd Farris of Stanolind Oil and Gas Corporation. This study became a basis of the first hydraulic fracturing experiment, which was conducted in 1947 at the Hugoton gas field in Grant County of southwestern Kansas by Stanolind. For the well treatment 1,000 US gallons (3,800 l; 830 imp gal) of gelled gasoline (essentially napalm) and sand from the Arkansas River was injected into the gas-producing limestone formation at 2,400 feet (730 m).

The experiment was not very successful as deliverability of the well did not change appreciably. The process was further described by J.B. Clark of Stanolind in his paper published in 1948. A patent on this process was issued in 1949 and an exclusive license was granted to the Halliburton Oil Well Cementing Company. On March 17, 1949, Halliburton performed the first two commercial hydraulic fracturing treatments in Stephens County, Oklahoma, and Archer County, Texas. Since then, hydraulic fracturing has been used to stimulate approximately a million oil and gas wells in various geologic regimes with good success.

In the Soviet Union, the first hydraulic proppant fracturing was carried out in 1952. Other countries in Europe and Northern Africa to use hydraulic fracturing included Norway, Poland, Czechoslovakia, Yugoslavia, Hungary, Austria, France, Italy, Bulgaria, Romania, Turkey, Tunisia, and Algeria.

Massive Hydraulic Fracturing

Figure: *Well Head where fluids are injected into the ground*

Figure: *Well head after all the Fracking equipment has been taken off location*

Pan American Petroleum applied the first massive hydraulic fracturing (also known as high-volume hydraulic fracturing) treatment in Stephens County, Oklahoma, USA in 1968. The definition of massive hydraulic fracturing varies somewhat, but is generally used for treatments injecting greater than about 150 mt, or approximately 330,000 pounds, of proppant.

American geologists became increasingly aware that there were huge volumes of gas-saturated sandstones with permeability too low (generally less than 0.1 millidarcy) to recover the gas economically. Starting in 1973, massive hydraulic fracturing was used in thousands of gas wells in the San Juan Basin, Denver Basin, the Piceance Basin, and the Green River Basin, and in other hard rock formations of the western US. Other tight sandstones in the US made economic by massive hydraulic fracturing were the Clinton-Medina Sandstone, and Cotton Valley Sandstone.

Massive hydraulic fracturing quickly spread in the late 1970s to western Canada, Rotliegend and Carboniferous gas-bearing sandstones in Germany, Netherlands onshore and offshore gas fields, and the United Kingdom sector of the North Sea. Horizontal oil or gas wells were unusual until the 1980s. Then in the late 1980s, operators in Texas began completing thousands of oil wells by drilling horizontally in the Austin Chalk, and giving massive slickwater hydraulic fracturing treatments to the wellbores. Horizontal wells proved much more effective than vertical wells in producing oil from the tight chalk; the shale runs horizontally so a horizontal well reached much more of the resource. In 1991, the first horizontal well was drilled in the Barnett Shale and in 1996 slickwater fluids were introduced.

Massive Hydraulic Fracturing in Shales

Due to shale's high porosity and low permeability, technology research, development and demonstration were necessary before hydraulic fracturing could be commercially applied to shale gas deposits. In the 1970s the United States government initiated the Eastern Gas Shales Project, a set of dozens of public-private hydraulic fracturing pilot demonstration projects. During the same period, the Gas Research Institute, a gas industry research consortium, received approval for research and funding from the Federal Energy Regulatory Commission.

In 1997, based on earlier techniques used by Union Pacific Resources, now part of Anadarko Petroleum Corporation, Mitchell Energy, now part of Devon Energy, developed the hydraulic fracturing technique known as "slickwater fracturing" which involves adding chemicals to water to increase the fluid flow, that made the shale gas extraction economical.

As of 2013, in addition to the United States several countries are planning to use hydraulic fracturing for unconventional oil and gas production.

Induced Hydraulic Fracturing

According to the United States Environmental Protection Agency (EPA) *hydraulic fracturing* is a process to stimulate a natural gas, oil, or geothermal energy well to maximize the extraction. The broader process, however, is defined by EPA as including the acquisition of source water, well construction, well stimulation, and waste disposal.

Uses

The technique of hydraulic fracturing is used to increase the rate at which fluids, such as petroleum, water, or natural gas can be recovered from subterranean natural reservoirs. Reservoirs are typically porous sandstones, limestones or dolomite rocks, but also include "unconventional reservoirs" such as shale rock or coal beds. Hydraulic fracturing enables the production of natural gas and oil from rock formations deep below the earth's surface (generally 5,000–20,000 feet (1,500–6,100 m)), which is typically greatly below groundwater reservoirs of basins if present.

At such depth, there may not be sufficient permeability or reservoir pressure to allow natural gas and oil to flow from the rock into the wellbore at economic rates. Thus, creating conductive fractures in the rock is pivotal to extract gas from shale reservoirs because of the extremely low natural permeability of shale, which is measured in the microdarcy to nanodarcy range. Fractures provide a conductive path connecting a larger volume of the reservoir to the well. So-called "super fracing", which creates cracks deeper in the rock formation to release more oil and gas, will increase efficiency of hydraulic fracturing. The yield for a typical shale gas well generally falls off after the first year or two, although the full producing life of a well can last several decades.

While the main industrial use of hydraulic fracturing is in arousing production from oil and gas wells, hydraulic fracturing is also applied:

- To stimulate groundwater wells
- To precondition or induce rock to cave in mining
- As a means of enhancing waste remediation processes, usually hydrocarbon waste or spills
- To dispose of waste by injection into deep rock formations

- As a method to measure the stress in the Earth
- For heat extraction to produce electricity in enhanced geothermal systems
- To increase injection rates for geologic sequestration of CO_2

Hydraulic Fracturing of Water-supply Wells

Since the late 1970s, hydraulic fracturing has been used in some cases to increase the yield of drinking water from wells in a number of countries, including the US, Australia, and South Africa.

Method

A hydraulic fracture is formed by pumping the fracturing fluid into the wellbore at a rate sufficient to increase pressure downhole at the target zone (determined by the location of the well casing perforations) to exceed that of the fracture gradient (pressure gradient) of the rock. The fracture gradient is defined as the pressure increase per unit of the depth due to its density and it is usually measured in pounds per square inch per foot or bars per metre. The rock cracks and the fracture fluid continues further into the rock, extending the crack still further, and so on. Fractures are localized because pressure drop off with frictional loss attributed to the distance from the well. Operators typically try to maintain "fracture width", or slow its decline, following treatment by introducing into the injected fluid a proppant – a material such as grains of sand, ceramic, or other particulates, that prevent the fractures from closing when the injection is stopped and the pressure of the fluid is removed. Consideration of proppant strengths and prevention of proppant failure becomes more important at greater depths where pressure and stresses on fractures are higher. The propped fracture is permeable enough to allow the flow of formation fluids to the well. Formation fluids include gas, oil, salt water and fluids introduced to the formation during completion of the well during fracturing.

During the process, fracturing fluid leakoff (loss of fracturing fluid from the fracture channel into the surrounding permeable rock) occurs. If not controlled properly, it can exceed 70% of the injected volume. This may result in formation matrix damage, adverse formation fluid interactions, or altered fracture geometry and thereby decreased production efficiency.

The location of one or more fractures along the length of the borehole is strictly controlled by various methods that create or seal off holes in the side of the wellbore. Hydraulic fracturing is performed in cased wellbores and the zones to be fractured are accessed by perforating the casing at those locations.

Hydraulic-fracturing equipment used in oil and natural gas fields usually consists of a slurry blender, one or more high-pressure, high-volume fracturing pumps (typically powerful triplex or quintuplex pumps) and a monitoring unit.

Associated equipment includes fracturing tanks, one or more units for storage and handling of proppant, high-pressure treating iron, a chemical additive unit (used to accurately monitor chemical addition), low-pressure flexible hoses, and many gauges and metres for flow rate, fluid density, and treating pressure. Chemical additives are typically 0.5% percent of the total fluid volume. Fracturing equipment operates over a range of pressures and injection rates, and can reach up to 100 megapascals (15,000 psi) and 265 litres per second (9.4 cu ft/s) (100 barrels per minute).

Well Types

A distinction can be made between conventional or low-volume hydraulic fracturing used to stimulate high-permeability reservoirs to frac a single well, and unconventional or high-volume hydraulic fracturing, used in the completion of tight gas and shale gas wells as unconventional wells are deeper and require higher pressures than conventional vertical wells. In addition to hydraulic fracturing of vertical wells, it is also performed in horizontal wells. When done in already highly permeable reservoirs such as sandstone-based wells, the technique is known as "well stimulation".

Horizontal drilling involves wellbores where the terminal drillhole is completed as a "lateral" that extends parallel with the rock layer containing the substance to be extracted. For example, laterals extend 1,500 to 5,000 feet (460 to 1,500 m) in the Barnett Shale basin in Texas, and up to 10,000 feet (3,000 m) in the Bakken formation in North Dakota. In contrast, a vertical well only accesses the thickness of the rock layer, typically 50–300 feet (15–91 m). Horizontal drilling also reduces surface disruptions as fewer wells are required to access a given volume of reservoir rock.

Drilling usually induces damage to the pore space at the wellbore wall, reducing the permeability at and near the wellbore. This reduces flow into the borehole from the surrounding rock formation, and partially seals off the borehole from the surrounding rock. Hydraulic fracturing can be used to restore permeability, but is not typically administered in this way.

Fracturing Fluids

Figure: *Water tanks preparing for a frac job*

High-pressure fracture fluid is injected into the wellbore, with the pressure above the fracture gradient of the rock. The two main purposes of fracturing fluid is to extend fractures, add lubrication, change gel strength and to carry proppant into the formation, the purpose of which is to stay there without damaging the formation or production of the well. Two methods of transporting the proppant in the fluid are used – high-rate and high-viscosity. High-viscosity fracturing tends to cause large dominant fractures, while high-rate (slickwater) fracturing causes small spread-out micro-fractures.

This fracture fluid contains water-soluble gelling agents (such as guar gum) which increase viscosity and efficiently deliver the proppant into the formation.

The fluid injected into the rock is typically a slurry of water, proppants, and chemical additives. Additionally, gels, foams, and compressed gases, including nitrogen, carbon dioxide and air can be injected. Typically, of the fracturing fluid 90% is water and 9.5% is sand with the chemical additives accounting to about 0.5%. However, fracturing fluids have been developed in which the use of water has been made unnecessary, using liquefied petroleum gas (LPG) and propane.

Figure: *Process of mixing water with fracking fluids to be injected into the ground*

A proppant is a material that will keep an induced hydraulic fracture open, during or following a fracturing treatment, and can be gel, foam, or slickwater-based. Fluids make tradeoffs in such material properties as viscosity, where more viscous fluids can carry more concentrated proppant; the energy or pressure demands to maintain a certain flux pump rate (flow velocity) that will conduct the proppant appropriately; pH, various rheological factors, among others.

Types of proppant include silica sand, resin-coated sand, and man-made ceramics. These vary depending on the type of permeability or grain strength needed. The most commonly used proppant is silica sand, though proppants of uniform size and shape, such as a ceramic proppant, is believed to be more effective. Due to a higher porosity within the fracture, a greater amount of oil and natural gas is liberated.

The fracturing fluid varies in composition depending on the type of fracturing used, the conditions of the specific well being fractured, and the water characteristics. A typical fracture treatment uses between 3 and 12 additive chemicals. Although there may be unconventional fracturing fluids, the more typically used chemical additives can include one or more of the following:

- Acids—hydrochloric acid (usually 5%-28%), or acetic acid is used in the pre-fracturing stage for cleaning the perforations and initiating fissure in the near-wellbore rock.
- Sodium chloride (salt)—delays breakdown of the gel polymer chains.
- Polyacrylamide and other friction reducers—minimizes the friction between fluid and pipe, thus allowing the pumps to pump at a higher rate without having greater pressure on the surface.
- Ethylene glycol—prevents formation of scale deposits in the pipe.
- Borate salts—used for maintaining fluid viscosity during the temperature increase.
- Sodium and potassium carbonates—used for maintaining effectiveness of crosslinkers.
- Glutaraldehyde—used as disinfectant of the water (bacteria elimination).
- Guar gum and other water-soluble gelling agents—increases viscosity of the fracturing fluid to deliver more efficiently the proppant into the formation.
- Citric acid—used for corrosion prevention.
- Isopropanol—increases the viscosity of the fracture fluid.

The most common chemical used for hydraulic fracturing in the United States in 2005–2009 was methanol, while some other most widely used chemicals were isopropyl alcohol, 2-butoxyethanol, and ethylene glycol.

Typical fluid types are:

- Conventional linear gels. These gels are cellulose derivatives (carboxymethyl cellulose, hydroxyethyl cellulose, carboxymethyl hydroxyethyl cellulose, hydroxypropyl cellulose, methyl hydroxyl ethyl cellulose), guar or its derivatives (hydroxypropyl guar, carboxymethyl hydroxypropyl guar) based, with other chemicals providing the necessary chemistry for the desired results.
- Borate-crosslinked fluids. These are guar-based fluids cross-linked with boron ions (from aqueous borax/boric acid solution). These gels have higher viscosity at pH 9 onwards and are used to carry proppants. After the fracturing job the pH is reduced

to 3–4 so that the cross-links are broken and the gel is less viscous and can be pumped out.

- Organometallic-crosslinked fluids zirconium, chromium, antimony, titanium salts are known to crosslink the guar based gels. The crosslinking mechanism is not reversible. So once the proppant is pumped down along with the cross-linked gel, the fracturing part is done. The gels are broken down with appropriate breakers.
- Aluminium phosphate-ester oil gels. Aluminium phosphate and ester oils are slurried to form cross-linked gel. These are one of the first known gelling systems.

For slickwater it is common to include sweeps or a reduction in the proppant concentration temporarily to ensure the well is not overwhelmed with proppant causing a screen-off. As the fracturing process proceeds, viscosity reducing agents such as oxidizers and enzyme breakers are sometimes then added to the fracturing fluid to deactivate the gelling agents and encourage flowback. The oxidizer reacts with the gel to break it down, reducing the fluid's viscosity and ensuring that no proppant is pulled from the formation. An enzyme acts as a catalyst for the breaking down of the gel. Sometimes pH modifiers are used to break down the crosslink at the end of a hydraulic fracturing job, since many require a pH buffer system to stay viscous. At the end of the job the well is commonly flushed with water (sometimes blended with a friction reducing chemical) under pressure. Injected fluid is to some degree recovered and is managed by several methods, such as underground injection control, treatment and discharge, recycling, or temporary storage in pits or containers while new technology is continually being developed and improved to better handle waste water and improve re-usability.

Fracture Monitoring

Measurements of the pressure and rate during the growth of a hydraulic fracture, as well as knowing the properties of the fluid and proppant being injected into the well provides the most common and simplest method of monitoring a hydraulic fracture treatment. This data, along with knowledge of the underground geology can be used to model information such as length, width and conductivity of a propped fracture.

Injection of radioactive tracers, along with the other substances in hydraulic-fracturing fluid, is sometimes used to determine the injection profile and location of fractures created by hydraulic fracturing. The radiotracer is chosen to have the readily detectable radiation, appropriate chemical properties, and a half life and toxicity level that will minimize initial and residual contamination. Radioactive isotopes chemically bonded to glass (sand) and/or resin beads may also be injected to track fractures. For example, plastic pellets coated with 10 GBq of Ag-110mm may be added to the proppant or sand may be labelled with Ir-192 so that the proppant's progress can be monitored. Radiotracers such as Tc-99m and I-131 are also used to measure flow rates. The Nuclear Regulatory Commission publishes guidelines which list a wide range of radioactive materials in solid, liquid and gaseous forms that may be used as tracers and limit the amount that may be used per injection and per well of each radionuclide.

Microseismic Monitoring

For more advanced applications, microseismic monitoring is sometimes used to estimate the size and orientation of hydraulically induced fractures. Microseismic activity is measured by placing an array of geophones in a nearby wellbore. By mapping the location of any small seismic events associated with the growing hydraulic fracture, the approximate geometry of the fracture is inferred. Tiltmeter arrays, deployed on the surface or down a well, provide another technology for monitoring the strains produced by hydraulic fracturing.

Microseismic mapping is very similar geophysically seismology. In earthquake seismology seismometers scattered on or near the surface of the earth record S-waves and P-waves that are released during an earthquake event. This allows for the motion along the fault plane to be estimated and its location in the earth's subsurface mapped. During formation stimulation by hydraulic fracturing an increase in the formation stress proportional to the net fracturing pressure as well as an increase in pore pressure due to leakoff takes place. Tensile stresses are generated ahead of the fracture/cracks' tip which generates large amounts of shear stress. The increase in pore water pressure and formation stress combine and affect the weakness (natural fractures, joints, and bedding planes) near the hydraulic fracture. Dilatational and compressive reactions occur and emit seismic energy detectable by highly sensitive geophones placed in nearby wells or on the surface.

Different methods have different location errors and advantages. Accuracy of microseismic event locations is dependent on the signal to noise ratio and the distribution of the receiving sensors. For a surface array location accuracy of events located by seismic inversion is improved by sensors placed in multiple azimuths from the monitored borehole. In a downhole array location accuracy of events is improved by being close to the monitored borehole (high signal to noise ratio).

Monitoring of microseismic events induced by reservoir stimulation has become a key aspect in evaluation of hydraulic fractures and their optimization. The main goal of hydraulic fracture monitoring is to completely characterize the induced fracture structure and distribution of conductivity within a formation. This is done by first understanding the fracture structure. Geomechanical analysis, such as understanding the material properties, in-situ conditions and geometries involved will help with this by providing a better definition of the environment in which the hydraulic fracture network propagates. The next task is to know the location of proppant within the induced fracture and the distribution of fracture conductivity. This can be done using multiple types of techniques and finally, further develop a reservoir model than can accurately predict well performance.

Horizontal Completions

Since the early 2000s, advances in drilling and completion technology have made drilling horizontal wellbores much more economical. Horizontal wellbores allow for far greater exposure to a formation than a conventional vertical wellbore. This is particularly useful in shale formations which do not have sufficient permeability to produce economically with a vertical well. Such wells when drilled onshore are now usually hydraulically fractured in a number of stages, especially in North America. The type of wellbore completion used will affect how many times the formation is fractured, and at what locations along the horizontal section of the wellbore.

In North America, shale reservoirs such as the Bakken, Barnett, Monterey, Haynesville, Marcellus, and most recently the Eagle Ford, Niobrara and Utica shales are drilled, completed and fractured using this method. The method by which the fractures are placed along the wellbore is most commonly achieved by one of two methods, known as "plug and perf" and "sliding sleeve".

The wellbore for a plug and perf job is generally composed of standard joints of steel casing, either cemented or uncemented, which is set in place at the conclusion of the drilling process. Once the drilling rig has been removed, a wireline truck is used to perforate near the end of the well, following which a fracturing job is pumped (commonly called a stage). Once the stage is finished, the wireline truck will set a plug in the well to temporarily seal off that section, and then perforate the next section of the wellbore. Another stage is then pumped, and the process is repeated as necessary along the entire length of the horizontal part of the wellbore.

The wellbore for the sliding sleeve technique is different in that the sliding sleeves are included at set spacings in the steel casing at the time it is set in place. The sliding sleeves are usually all closed at this time. When the well is ready to be fractured, using one of several activation techniques, the bottom sliding sleeve is opened and the first stage gets pumped. Once finished, the next sleeve is opened which concurrently isolates the first stage, and the process repeats. For the sliding sleeve method, wireline is usually not required.

Figure: *Sleeves*

These completion techniques may allow for more than 30 stages to be pumped into the horizontal section of a single well if required, which is far more than would typically be pumped into a vertical well.

Economic Impacts

Hydraulic fracturing has been seen as one of the key methods of extracting unconventional oil and gas resources. According to the International Energy Agency, the remaining technically recoverable resources of shale gas are estimated to amount to 208 trillion cubic metres (208,000 km^3), tight gas to 76 trillion cubic metres (76,000 km^3), and coalbed methane to 47 trillion cubic metres (47,000 km^3). As a rule, formations of these resources have lower permeability than conventional gas formations. Therefore, depending on the geological characteristics of the formation, specific technologies (such as hydraulic fracturing) are required. Although there are also other methods to extract these resources, such as conventional drilling or horizontal drilling, hydraulic fracturing is one of the key methods making their extraction economically viable. The multi-stage fracturing technique has facilitated the development of shale gas and light tight oil production in the United States and is believed to do so in the other countries with unconventional hydrocarbon resources.

The National Petroleum Council estimates that hydraulic fracturing will eventually account for nearly 70% of natural gas development in North America. Hydraulic fracturing and horizontal drilling apply the latest technologies and make it commercially viable to recover shale gas and oil. In the United States, 45% of domestic natural gas production and 17% of oil production would be lost within 5 years without usage of hydraulic fracturing.

A number of studies related to the economy and fracking, demonstrates a direct benefit to economies from fracking activities in the form of personnel, support, ancillary businesses, analysis and monitoring. Typically the funding source of the study is a focal point of controversy. Most studies are either funded by mining companies or funded by environmental groups, which can at times lead to at least the appearance of unreliable studies. An unbiased study was performed by Deller & Schreiber in 2012, looking at the relationship between non-oil and gas mining and community economic growth. The study concluded that there is an impact on income growth; however, researchers found that mining does not lead to an increase in population or employment. The actual financial impact of non-oil and gas mining on the economy is dependent on many variables and is difficult to identify definitively.

Environmental Impact

Hydraulic fracturing has raised environmental concerns and is challenging the adequacy of existing regulatory regimes. These concerns have included ground water contamination, risks to air quality, migration of gases and hydraulic fracturing chemicals to the surface, mishandling of waste, and the health effects of all these, as well as its contribution to raised atmospheric CO_2 levels by enabling the extraction of previously-sequestered hydrocarbons. Because hydraulic fracturing originated in the United States, its history is more extensive there than in other regions. Most environmental impact studies have therefore taken place there.

Research Issues

Several organizations, researchers, and media outlets have reported difficulty in conducting and reporting the results of studies on hydraulic fracturing due to industry and governmental pressure, and expressed concern over possible censoring of environmental reports., though work by National Science Foundation, the EPA and several universities has been considered unbiased. Concerns have been raised about the role of wealthy foundations in financing research that some have argued was designed inflate the risks of development, and lobbying by the gas industry to promote its activities. The broader debate over these topics provides an example of the research challenges on this subject. Researchers have recommended requiring disclosure of all hydraulic fracturing fluids, testing animals raised near fracturing sites, and closer monitoring of environmental samples. After court cases concerning contamination from hydraulic fracturing are settled, the documents are sealed, and at least one recent case bears that out, while others believe it has and could lead to unnecessary risks to public safety and health. The American Petroleum Institute denies that this practice has hidden problems with gas drilling.

Air

The air emissions from hydraulic fracturing are related to methane leaks originating from wells, and emissions from the diesel or natural gas powered equipment such as compressors, drilling rigs, pumps etc. Also transportation of necessary water volume for hydraulic fracturing, if done by trucks, can cause high volumes of air emissions, especially particulate matter emissions. There are also reports of health problems around compressors stations or drilling sites, although a causal relationship is not established and one Texas analysis found no evidence of effects.

Natural gas produced by hydraulic fracturing causes higher well-to-burner emissions than gas produced from conventional wells. Although a recent report coauthored by researchers at the U.S. Department of Energy's National Renewable Energy Laboratory found emissions from shale gas, when burned for electricity, were "very similar" to those from so-called "conventional well" natural gas, hydraulic fracturing's higher emissions profile is mainly due to the gas released during completing wells as some gas returns to the surface, together with the fracturing fluids. Depending on their treatment, the well-to-burner emissions are 3.5%–12% higher than for conventional gas. Other studies have found different effects, and a debate has arisen particularly around a study by professor Robert W. Howarth finding shale gas significantly worse for global warming than oil or coal and various others criticizing the analysis. Howarth has responded that "The latest EPA estimate for methane emissions from shale gas falls within the range of our estimates but not those of Cathles *et al.*, which are substantially lower."

The U.S. EPA has estimated the methane leakage rate to be about 2.4% – well below Howarth's estimate. The American Gas Association, and industry trade group, calculated a 1.2% leakage rate based on the EPA's latest greenhouse gas inventory, although the EPA has not publicly stated a change to its prior estimate.

Water

Consumption: The large volumes of water required have raised concerns about hydraulic fracturing in arid areas, such as Karoo in South Africa and drought prone areas of North America. During periods of low stream flow it may affect water supplies for municipalities and industries such as power generation, as well as recreation and aquatic life. It may also require water overland piping from distant sources.

Hydraulic fracturing uses between 1.2 and 3.5 million US gallons (4.5 and 13 Ml) of water per well, with large projects using up to 5 million US gallons (19 Ml). Additional water is used when wells are refractured. An average well requires 3 to 8 million US gallons (11,000 to 30,000 m^3) of water over its lifetime. Using the case of the Marcellus Shale as an example, as of 2008 hydraulic fracturing accounted for 650 million US gallons per year (2,500,000 m^3/a) or less than 0.8% of annual water use in the area overlying the Marcellus Shale. The annual number of well permits, however, increased by a factor of five and the number of well starts increased by a factor of over 17 from 2008 to 2011. According to the Oxford Institute for Energy Studies, greater volumes of fracturing fluids are required in Europe, where the shale depths average 1.5 times greater than in the U.S. To minimize water consumption, recycling is one possible option. In the Spring of 2013, new hydraulic fracturing water recycling rules were adopted in the state of Texas by the Railroad Commission of Texas. The Water Recycling Rules are intended to encourage Texas hydraulic fracturing operators to conserve water used in the hydraulic fracturing process for oil and gas wells. Another possible option is to use carbon dioxide instead of water.

Injected Fluid

There are concerns about possible contamination by hydraulic fracturing fluid both as it is injected under high pressure into the ground and as it returns to the surface. To mitigate the impact of hydraulic fracturing to groundwater, the well and ideally the shale formation itself should remain hydraulically isolated from other geological formations, especially freshwater aquifers. In 2009 state regulators from at least a dozen states have also stated that they have seen no evidence of the hydraulic fracturing process polluting drinking water. In May 2011, former U.S. EPA administrator Lisa Jackson (appointed by President Barack Obama) has said on at least two occasions that there is either no proven case of direct contamination by the hydraulic fracturing process, or that the EPA has never made a definitive determination of such contamination. By August 2011 there were at least 36 cases of suspected groundwater contamination due to hydraulic fracturing in the United States. In more recent congressional testimony in April 2013, Dr. Robin Ikeda, Deputy Director of Noncommunicable Diseases, Injury and Environmental Health at the CDC listed several sites where EPA had documented contamination. In several cases EPA has determined that hydraulic fracturing was likely the source of the contamination.

While some of the chemicals used in hydraulic fracturing are common and generally harmless, some are known carcinogens at high enough doses. A report prepared for House Democratic members Henry Waxman, Edward Markey and Diana DeGette stated that out of 2,500 hydraulic fracturing products, "more than 650 of these products contained chemicals that are known or possible human carcinogens, regulated under the Safe Drinking Water Act, or listed as hazardous air pollutants". The report also shows that between 2005 and 2009, 279 products had at least one component listed as "proprietary" or "trade secret" on their Occupational Safety and Health Administration (OSHA) required material safety data sheet (MSDS). The MSDS is a list of chemical components in the products of chemical manufacturers, and according to OSHA, a manufacturer may withhold information designated as "proprietary" from this sheet. When asked to reveal the proprietary components, most companies participating in the investigation were unable to do so, leading the committee to surmise these "companies are injecting fluids containing unknown chemicals about which they may have limited understanding of the potential risks posed to human health and the environment". Without knowing the identity of the proprietary components, regulators cannot test for their presence. This prevents government regulators from establishing baseline levels of the substances prior to hydraulic fracturing and documenting changes in these levels, thereby making it more difficult to prove that hydraulic fracturing is contaminating the environment with these substances.

Another 2011 study identified 632 chemicals used in natural gas operations. Only 353 of these are well-described in the scientific literature. The study indicated possible long-term health effects that might not appear immediately. The study recommended full disclosure of all products used, along with extensive air and water monitoring near natural gas operations; it also recommended that hydraulic fracturing's exemption from regulation under the US Safe Drinking Water Act be rescinded. Industry group *Energy In Depth*, a research arm of the Independent Petroleum Association of America, contends that fracking "was never granted an 'exemption' from it... How can something earn an exemption from a law that never covered or even conceived of it in the first place?"

Governments are responding to questions about the contents of hydraulic fracturing fluid by requiring disclosure via government agencies and public web site. The Irish regulatory regime requires full disclosure of all additives to Ireland's Environmental Protection Agency (Ireland). The European Union also requires such disclosure. In the US, the Ground Water Protection Council launched FracFocus.org, an online voluntary disclosure database for hydraulic fracturing fluids funded by oil and gas trade groups and the U.S. Department of Energy. The site has been met with some skepticism relating to proprietary information that is not included. Some states have mandated fluid disclosure and incorporated FracFocus as the tool for disclosure. Also in the US, FracTracker Alliance provides oil and gas-related data storage, analyses, and online and customized maps related to hydraulic fracturing on FracTracker.org.

Flowback

As the fracturing fluid flows back through the well, it consists of spent fluids and may contain dissolved constituents such as minerals and brine waters. It may account for about 30–70% of the original fracture fluid volume. In addition, natural formation waters may flow to the well and need treatment or disposal. These fluids, commonly known as flowback, produced water, or wastewater, are managed by underground injection, wastewater treatment and discharge, or recycling to fracture future wells. Hydraulic fracturing can concentrate levels of uranium, radium, radon, and thorium in flowback. Treatment of produced waters may be feasible through either self-contained systems at well sites or fields or through municipal waste water treatment plants or commercial treatment facilities. However, the quantity of waste water being treated, and the improper configuration of sewage plants to treat it, have become an issue in the northeast United States. Much of the wastewater from hydraulic fracturing operations in Pennsylvania is processed by public sewage treatment plants, which are not equipped to remove radioactive material and are not required to test for it. Another issue is the bromide in waste brine. The bromide combines with chlorine disinfectant and dissolved organic matter in water treatment plants to form trihalomethanes (THMs).

Interestingly, a recent study from Duke University found: "Contrary to current perceptions, Marcellus [Shale] wells produce significantly less wastewater per unit gas recovered (~35%) compared to conventional natural gas wells." Estimates of the amount of injected fluid returning to the surface vary. Some say approximately 15-20% of the injected fluid returns to the surface with the gas and other that higher amounts return. Some remains underground and some may return to the surface through abandoned wells or other pathways. Although not necessarily indicative of broader industry trends, several reports have also highlighted an industry-wide shift toward greater water recycling in the Marcellus Shale.

Methane

Groundwater methane contamination is also a concern as it has adverse impact on water quality and in extreme cases may lead to potential explosion. In 2006, over 7 million cubic feet (200,000 m^3) of methane were released from a blown gas well in Clark, Wyoming and shallow groundwater was found to be contaminated. However, methane contamination is not always caused by hydraulic fracturing. Drilling for ordinary drinking water wells can also cause methane release. Some studies make use of tests that can distinguish between the deep thermogenic methane released during gas/oil drilling, and the shallower biogenic methane that can be released during water-well drilling. While both forms of methane result from decomposition, thermogenic methane results from geothermal assistance deeper underground.

According to the 2011 study of the MIT Energy Initiative, "there is evidence of natural gas (methane) migration into freshwater zones in some areas, most likely as a result of substandard well completion practices i.e. poor quality cementing job or bad casing, by a few operators." 2011 studies by the Colorado School of Public Health and Duke University also pointed to methane contamination stemming from hydraulic fracturing or its surrounding process. A study by Cabot Oil and Gas examined the Duke study using a larger sample size, found that methane concentrations were related to topography, with the highest readings found in low-lying areas, rather than related to distance from gas production areas. Using a more precise isotopic analysis, they showed that the methane found in the water wells came from both the Marcellus Shale (Middle Devonian) where hydraulic fracturing occurred, and from the shallower Upper Devonian formations. A 2013 Duke study suggested that both defective cement seals in the upper part of wells and faulty steel linings within deeper layers may be allowing methane and injected fluid to seep into surface waters. Abandoned gas and oil wells also provide conduits to the surface.

In April 2013 the EPA dramatically lowered its estimate of how much methane gas is released to the atmosphere during the fracking process by 20 percent.

Radioactivity

There are concerns about the levels of radioactivity in wastewater from hydraulic fracturing and its potential impact on public health. A Popular Mechanics article stated, however, that although shale does have a radioactive signature, testing conducted in Pennsylvania in 2009 found "no evidence of elevated radiation levels" in waterways. The EPA called for more testing. In 2009, Conrad Dan Volz, former Director of the Centre for Health Environments and Communities at the University of Pittsburgh, said that radiation concerns are one of the least pressing issues.

In 2011 *The New York Times* reported radium in wastewater from natural gas wells is released into Pennsylvania rivers, and has compiled a map of these wells and their wastewater contamination levels, and stated that some EPA reports were never made public. The *Times'* reporting on the issue has come under some criticism. Recycling this wastewater has been proposed as a partial solution, but this approach has limitations. A 2012 study examining a number of hydraulic fracturing sites in Pennsylvania and Virginia by Pennsylvania State University, found that water that flows back from gas wells after hydraulic fracturing contains high levels of radium. Solid waste such as drill clippings is also radioactive. In 2012 there were 1325 radiation alerts from all sources at dumps in Pennsylvania, up from 423 alerts in 2008. At least 1,000 of the 2012 alerts were set off by waste from gas and oil drilling hydraulic fracturing operations.

Seismicity

Hydraulic fracturing routinely produces microseismic events much too small to be detected except by sensitive instruments. These microseismic events are often used to map the horizontal and vertical extent of the fracturing. However, as of late 2012, there have been three instances of hydraulic fracturing, through induced seismicity, triggering quakes large enough to be felt by people: one each in the United States, Canada, and England.

The injection of waste water from oil and gas operations, including from hydraulic fracturing, into saltwater disposal wells may cause bigger low-magnitude tremors, being registered up to 3.3 (M_w).

Induced Seismicity from Hydraulic Fracturing

The United States Geological Survey (USGS) has reported earthquakes induced by hydraulic fracturing, and by disposal of hydraulic fracturing flowback into waste disposal wells, in several locations. Bill Ellsworth, a geoscientist with the U.S. Geological Survey, has said, however: "We don't see any connection between fracking and earthquakes of any concern to society." The National Research Council (part of the National Academy of Sciences) has also observed that hydraulic fracturing, when used in shale gas recovery, does not pose a serious risk of causing earthquakes that can be felt.

A British Columbia Oil and Gas Commission investigation concluded that a series of 38 earthquakes (magnitudes ranging from 2.2 to 3.8 on the Richter scale) occurring in the Horn River Basin area between 2009 and 2011 were caused by fluid injection during hydraulic fracturing in proximity to pre-existing faults. The tremors were small enough that only one of them was reported felt by people; there were no reports of injury or property damage.

A report in the UK concluded that hydraulic fracturing was the likely cause of two small tremors (magnitudes 2.3 and 1.4 on the Richter scale) that occurred during hydraulic fracturing of shale.

Induced Seimicity from Water Disposal Wells

According to the USGS only a small fraction of roughly 40,000 waste fluid disposal wells for oil and gas operations in the United States have induced earthquakes that are large enough to be of concern to the public. Although the magnitudes of these quakes has been small, the USGS says that there is no guarantee that larger quakes will not occur. In addition, the frequency of the quakes has been increasing. In 2009, there were 50 earthquakes greater than magnitude 3.0 in the area spanning Alabama and Montana, and there were 87 quakes in 2010. In 2011 there were 134 earthquakes in the same area, a sixfold increase over 20th century levels. There are also concerns that quakes may damage underground gas, oil, and water lines and wells that were not designed to withstand earthquakes.

Several earthquakes in 2011, including a 4.0 magnitude quake on New Year's Eve that hit Youngstown, Ohio, are likely linked to a disposal of hydraulic fracturing wastewater, according to seismologists at Columbia University. A similar series of small earthquakes occurred in 2012 in Texas. Earthquakes are not common occurrences in either area. Disposal and injection wells are regulated under the Safe Drinking Water Act and UIC laws.

Health Impacts

Concern has been expressed over the possible long and short term health effects of air and water contamination and radiation exposure by gas production. A study on the effect of gas drilling, including hydraulic fracturing, published by the Cornell University College of Veterinary Medicine, concluded that exposure to gas drilling operations was strongly implicated in serious health effects on humans and animals although scientists have raised concerns about that particular report. As of May 2012, the United States Institute of Medicine and United States National Research Council were preparing to review the potential human and environmental risks of hydraulic fracturing.

The U.S. Environmental Protection Agency considers radioactive material in flowback a hazard to workers at hydraulic fracturing sites. Workers may inhale radon gas released by the process, raising their risk of lung cancer. They are also exposed to alpha and gamma radiation released during the decay of radium-226 and to gamma radiation and beta particles released by the decay of radium-228, according to EPA. EPA reports that gamma radiation can also penetrate the skin and raise the risk of cancer.

A 2012 study concluded that risk prevention efforts should be directed towards reducing air emission exposures for persons living and working near wells during well completions. In the United States the Occupational Safety and Health Administration (OSHA) and the National Institute for Occupational Safety and Health (NIOSH) released a hazard alert based on data collected by NIOSH that workers may be exposed to dust with high levels of respirable crystalline silica (silica dioxide) during hydraulic fracturing. NIOSH notified company representatives of these findings and provided reports with recommendations to control exposure to crystalline silica and recommend that all hydraulic fracturing sites evaluate their operations to determine the potential for worker exposure to crystalline silica and implement controls as necessary to protect workers.

According to the United States Department of Energy, hydraulic fracturing fluid is composed of approximately 95% water, 4.5% sand and 0.5% different chemicals. These percentages are by weight, so hydraulically fracturing a well uses 4-7 million gallons of water (15000-27000 tons) and 80-140 tons of chemicals. There can be up to 65 chemicals and often include benzyne, glycol-ethers, toluene, ethanol and nonphenols. Some have argued that although many of these chemicals are harmful, some of them are either non toxic or are non toxic at lower dosages. However, their concentration in hydraulic fracturing fluid have proven toxic to animals and humans. Many chemicals used in fracking, such as 2-BE ethylene glycol, are carcenogenic. This chemical is listed under chronic oral RFD assessment, chronic inhalation RFC assessment, and carcinogenicity assessment records of the US environmental protection agency's website. In a study done by the US Environmental Protection Agency, it found statistically significant effects observed in mice included forestomach ulcers and epithelial hyperplasia, hematopoietic cell proliferation and hemosiderin pigmentation in the spleen, Kupffer cell pigmentation in the livers, and bone marrow hyperplasia (in males only, suggesting tissue damage due to exposure above 125-250ppm. The study also found statistically significant decreases in automated and manual hematocrit (Hct) values, hemoglobin (Hb) concentrations, and red blood cell (RBC) for both males and females at exposure of 250ppm and for female in the 125ppm exposure group.

In a study done by Colborn and colleagues, they examined 353 out of 994 fracking chemicals identified by TEDX in hydraulic fracking operation. They found over 75% of the 353 chemicals affected the skin, eyes, and other sensory organs, 52% affected the nervous system, 40% affected the immune system and kidney system, and 46% affected the cardiocascular system and blood.

In a second study done by Colborn and colleagues, they examined the airborne chemicals due to the fracking process. The group categorized the human tissue types into 12 categories and found 35 chemicals affected the brain/nervous system, 33 the liver/ metabolism, and 30 the endocrine system, which includes reproductive and developmental effects. The categories with the next highest numbers of effects were the immune system (28), cardiovascular/blood (27), and the sensory and respiratory systems (25 each). Eight chemicals had health effects in all 12 categories.

Airborne chemicals during the fracking process, such as benzene and benzene derivatives, naphthalene, methylene chloride, are either carcinogenic or suspected as a human carcinogen to the human body.

Politics and Public Policy

To control the hydraulic fracturing industry, some governments are developing legislation and some municipalities are developing local zoning limitations. In 2011, France became the first nation to ban hydraulic fracturing. Some other countries have placed a temporary moratorium on the practice. The US has the longest history with hydraulic fracturing, so its approach to hydraulic fracturing may be modelled by other countries.

The considerable opposition against hydraulic fracturing activities in local townships has led companies to adopt a variety of public relations measures to assuage fears about hydraulic fracturing, including the admitted use of "mil-i-tary tac-tics to counter drilling oppo-nents". At a conference where public relations measures were discussed, a senior executive at Anadarko Petroleum was recorded on tape saying, "Download the US Army / Marine Corps Counterinsurgency Manual, because we are dealing with an insurgency", while referring to hydraulic fracturing opponents. Matt Pitzarella, spokesman for Range Resources also told other conference attendees that Range employed psychological warfare operations veterans. According to Pitzarella, the experience learned in the Middle East has been valuable to Range Resources in Pennsylvania, when dealing with emotionally charged township meetings and advising townships on zoning and local ordinances dealing with hydraulic fracturing.

Police officers have recently been forced, however, to deal with intentionally disruptive and even potentially violent opposition to oil and gas development. In March 2013, ten people were arrested during an "anti-fracking protest" near New Matamoras, Ohio, after they illegally entered a development zone and latched themselves to drilling equipment. In northwest Pennsylvania, there was a drive-by shooting at a well site, in which an individual shot two rounds of a small-caliber rifle in the direction of a drilling rig, just before shouting profanities at the site and fleeing the scene. And in Washington County, Pa., a contractor working on a gas pipeline found a pipe bomb that had been placed where a pipeline was to be constructed, which local authorities said would have caused a "catastrophe" had they not discovered and detonated it.

Media Coverage

Josh Fox's 2010 Academy Award nominated film *Gasland* became a centre of opposition to hydraulic fracturing of shale. The movie presented problems with ground water contamination near well sites in Pennsylvania, Wyoming, and Colorado. *Energy in Depth*, an oil and gas industry lobbying group, called the film's facts into question. In response, a rebuttal of *Energy in Depth*'s claims of inaccuracy was posted on *Gasland's* website. The Director of the Colorado Oil and Gas Conservation Commission (COGCC) offered to be interviewed as part of the film if he could review what was included from the interview in the final film but Fox declined the offer. Exxon Mobil, Chevron Corporation and ConocoPhillips aired advertisements during 2011 and 2012 that claim to describe the economic and environmental benefits of natural gas and argue hydraulic fracturing is safe.

The film *Promised Land*, starring Matt Damon, takes on hydraulic fracturing. The gas industry has made plans to counter the film's criticisms of hydraulic fracturing with informational flyers, and Twitter and Facebook posts.

One *New York Times* report claimed that an early draft of a 2004 EPA study discussed "possible evidence" of aquifer contamination but the final report omitted that mention. Some have criticized the narrowing of EPA studies, including the EPA study on hydraulic fracturing's impact on drinking water to be released in late 2014, such that hydrocarbon extraction processes not unique to hydraulic fracturing, such as drilling, casing, and above ground impacts, are considered beyond scope.

Chapter 2
Test Objectives

The primary purpose of performing a build-up test is to determine the wellbore damage (skin) and the reservoir permeability. However, during the course of a build-up, it is possible to encounter reservoir boundaries.

If all the reservoir's boundaries are contacted during the build-up, the size of the reservoir can also be determined. If the well has been pressure tested before, subsequent testing allows relative material balance calculations (decline curve analysis), as well as the determination of the drive mechanism for the reservoir.

Procedure

Install the SPIDR on a well that has been flowing steadily for several days. Check for leaks in the system after installation. The SPIDR must be recording for at least 15 minutes prior to shut in (check the box for the SPIDR wake-up time). Shut the well in manually at the wing. Do not shut in using the automatic choke?pit has a tendency to leak. After the well has been shut in, check for leaks again. When the build-up is over, the SPIDR may be rigged down and returned, or it may be left on the well for further testing while the well is flowing.

Advantages

A build-up test is one of the simplest tests to perform. The biggest advantage in performing a build-up is that it is a constant rate test: Q = 0. Simply install the equipment while the well is flowing, then shut the well in upstream of the choke.

Disadvantages

The drawback to performing a build-up test is that you lose cash flow. (Some people might think that you are actually losing gas, but it's still down there.) Nevertheless, if the test objectives are to determine skin and perm, and the rock has a permeability greater than 2 milliDarcies, most build-up tests can be limited to 2 days.

Commentary

It has been our experience that it is easier to obtain management approval for a build-up test when it is done in conjunction with a planned shut in due to facilities scheduled maintenance. After a build-up it is advisable to perform a single rate drawdown, or a Modified Isochronal test.

Drawdown Testing

Test Objectives

The objectives of a drawdown test are to determine skin, perm and the distances to the reservoir's boundaries. We recommend doing a drawdown test to look for reservoir limits instead of running a build-up because a flowing well test does not interrupt cash flow. Since you can't see a boundary until the pressure wave hits it, there's no way to tell how long a build-up is required to see the reservoir limits. On most tests, it is possible to perform an "Integrated Volume Explored" calculation as the test progresses. This allows for a test to be run until a given amount of reserves have been proven.

Procedure

Install the SPIDR on a well that has been shut-in and stable. "Stable" is defined as the shut-in well head pressure changing at a rate of less than 1 psi per hour. Check for leaks in the system after installation. The SPIDR must be recording for at least 15 minutes prior to opening the well (check the box for the SPIDR wake-up time). Begin flowing the well on a single choke size. If the well must be "stepped-up", try to get the well up to full rate within 30 minutes. Continue flowing on a constant choke size for the duration of the test. If shut-ins or flow interruptions occur during the course of the drawdown, try to get the well back on-line as soon as possible on the same choke size.

Advantages

The main benefit of running a drawdown is that cash flow is not interrupted. Another advantage is that reservoir boundaries are easier to locate, relative to build-up tests.

Disadvantages

The drawback to running a drawdown is that the rate may not be constant. However, changing the choke periodically to maintain a constant rate will cause more problems than letting the rate fluctuate. In order to get accurate analysis on a drawdown, it is critical that no choke changes occur during the test.

Computer-aided Analysis

Well Test Deconvolution: The deconvolution method has received much attention recently, and is becoming one of the major tools for well test and production data analysis in oil and gas industry. Here, we present a new deconvolution approach, which we believe is relevant and can be an important addition to the existing efforts made in this field. We show that the solution of the deconvolution problem can be successfully represented as a linear combination of non-orthogonal exponential functions. Also, we present three deconvolution algorithms. The first two algorithms are based on regularization concepts borrowed from the well-known Tikhonov and Krylov methods, while the third algorithm is based on the stochastic Monte Carlo method.

The current trends in well test analysis are motivated by the use of an improved gauge metrology, that allows continuous bottomhole well pressure monitoring. The increased flow of information from these permanent gauges should permit a better reservoir characterization, and the development of new proactive management and control techniques. The analysis and interpretation of the permanent gauge data is especially challenging because of the sensitivity of the involved methods to small perturbations and errors, and the size of the data sets, which may include thousands of flow periods at different rates and millions of pressure points collected over several recording years. Therefore, the development of robust computer-aided methods become crucial for modern well test analysis and interpretation.

Conventional well test analysis makes considerable use of graphical presentations of pressure derivatives, calculated over individual flow periods, which require a daunting task of identification and extraction from large amounts of data. In addition, the data is affected by errors, and can be easily distorted by the differentiation algorithms, thus providing an incorrect and misleading information. Also, the individual flow periods are of shorter durations compared to the entire test, which contains information about the reservoir at distances from the well which can be several orders of magnitude larger than the radius of investigation of a single flow period. Thus, the full information content from the measurements cannot be obtained with conventional analysis methods. The main objective of the modern deconvolution approach is to improve and extend the analysis beyond these restrictions, which are inherent in the traditional approach.

Deconvolution is a relatively new technique that can be used to convert the pressure data response from a variable-rate test, or a production sequence, into a constant-rate equivalent pressure profile for the entire duration of the test or production history, thus making the data more useful for interpretation. The deconvolution method extracts much more information from well test and production data than is possible by using conventional analysis methods. For example, the conventional Bourdet derivative plot displays the pressure behaviour only for a specific flow period of a test, while the result of deconvolution can give the transient pressure behaviour for a group of flow periods. Therefore, the deconvolved system response is defined for a longer time interval, and reveals the characteristics of the transient behaviour that otherwise would not be observed using the conventional methods.

The benefits of the deconvolution method are:

- it does not make any assumption on the reservoir type, and it does not require reservoir modelling;
- it provides information over a radius of investigation that can be orders of magnitude greater than that available from individual flow periods;
- it can reduce considerably the required duration of an extended well test, by revealing boundaries before they become visible in individual flow periods.

Applying deconvolution to well test and production data is a quite difficult computational challenge because it requires the solution of a large ill-conditioned problem. This ill-conditioned problem, combined with the errors that are inherent in the measurements and the large size of the data, makes the development of robust (error-tolerant) algorithms a real challenge. Recently, several deconvolution algorithms have been proposed. However, only a few of them appear to be somewhat more robust.[1-4]

Our deconvolution implementation includes some of the most robust algorithms developed so far:

- Nonlinear Total Least Squares (NTLS);[1,2]
- Markov Chain Monte Carlo (MCMC);[3]
- Redundant Dictionary Decomposition (RDD).[4]

The current and future development objectives of our approach to deconvolution are:

- to eliminate the subjective criteria for the selection of weight and regularization parameters;
- to develop efficient parallel code for the numerical solution of these problems, which can easily deal with large data sets;
- to formulate a mathematical and algorithmic framework for the analysis of well test data from multiple, possibly interfering wells, which is a common occurrence in high permeability reservoirs;
- to apply the algorithms to field data sets provided by the sponsors/investors.

The benefits of our deconvolution implementation approach are:

- cutting edge research in the theoretical and applied aspects of well test deconvolution;
- improvement of the best existing algorithms, and continuous addition of new significant algorithms;
- access to stable deconvolution algorithms which allow efficient interpretation of well test data from permanent downhole pressure gauges;
- accurate numerical results and fast computation.

What is A "Derivative Plot"

A derivative plot is a slightly different way to look at a traditional Horner plot. The big advantage of a derivative plot over the more traditional well-test analysis plots is that ALL the main well and reservoir characteristics are shown clearly on a single plot. This allows you to associate a pattern in the derivative plot to a particular well and reservoir description. I strongly recommend using the derivative plot; it really helps to improve a well-test analysis! However, don't get carried away; you still need to look at the traditional plots to do a complete analysis.

Here is an example of a Horner plot for a well with damage in a simple "infinite-acting" (i.e. no reservoir boundaries) system (time increases to the LEFT). Note the initial "S" shape for the well-bore storage effects followed by a straight-line.

A Derivative plot is displayed on log-log scales and is composed of two curves; an upper curve to plot delta-pressure verses delta-time, and a lower curve to plot the derivative verses delta-time. Why is this lower curve called the "derivative"? Because it is the first derivative of the Horner plot i.e. the instantaneous slope along the curve of the data on a Horner plot. Here is an example of the derivative plot; note how the well-bore storage "S" from the Horner plot becomes a "hump" with an initial UNIT SLOPE, and the straight-line on the Horner plot becomes a STABILISED derivative value.

And that is all there is to it. All of the other traditional plots (e.g. a square-root plot) will have a characteristic shape on the derivative plot (e.g. a half-slope).

"Basic Surveillance" Spreadsheet

This spreadsheet can be used to carry out well-test analysis calculations that are typically used for basic well performance surveillance and eliminates the need for a specialised software package (e.g. PIE) for this type of work. The key features in the spreadsheet are:

1. Use individual sheets for each set of shut-in data, with up to 32000 data points for each set of data and as many sheets as you want. A data "filter" is included to reduce the input pressure data to a manageable size.
2. A pseudo-pressure calculation is included which uses a table of gas viscosity and Z-factor data that you supply. The pseudo-pressure is used to transform the shut-in pressure data so the correct analysis results are obtained.
3. A Derivative plot for each shut-in is displayed along with a graphics control that is used to define a derivative stabilisation. This stabilisation is then used to estimate the reservoir permeability-thickness.
4. A Superposition plot for each shut-in is displayed along with a graphics control that is used to define a straight-line through the the data. This line is used to estimate the extrapolated reservoir pressure (P*) and the skin-factor.
5. Multiple sets of shut-in data can be compared on rate-normalised Derivative and Superposition plots (plot overlay). This comparison shows how well performance and reservoir pressure change over time.
6. The "Time to pseudo-steady-state flow" (Tpss) is estimated from a simple radius-of-investigation calculation, and a value of Tpss can be entered to control the length of the rate-history used in the superposition calculations.

All of the calculations are visible and unprotected so you can also use this spreadsheet to learn how basic well-test analysis calculations are carried out. All of the basic algorithms needed to prepare a derivative plot are included as Visual Basic code in the spreadsheet. This includes a general superposition function, and a simple derivative calculation which includes smoothing. Everything in the spreadsheet is open for you to modify as desired.

Gas Well Tests

Pseudo-Time: Pseudo-time is a mathematical time function that accounts for the variable compressibility(c_t) and viscosity (m_g) of gas as well as the variable total (formation) porosity (*f*) with respect to time and pressure.

Why Use Pseudo-time?

The equation for flow of gas in the reservoir is very similar to that for liquid flow. In well testing, analytical equations are solved after making certain assumptions. In particular, four assumptions are very important. These are:

- Total system compressibility (c_t) is constant
- Gas viscosity (μ_g) is constant
- Total porosity (ϕ) is constant
- Fluid saturations (S_g and S_w) are constant.

For liquids, these assumptions are reasonable, since liquid compressibility and viscosity do not vary significantly with pressure, and the equations can be solved analytically. These analytical solutions are referred to as the liquid flow solutions, and form the basis of all well test analysis. The result is an analytical relationship between pressure and time, which, for an infinite-acting reservoir, can be written as :

$$Pressure = Constant * \log (time) + ...$$

For gas, most of the assumptions listed above are no longer valid. Gas compressibility (c_g) varies significantly with pressure. Gas viscosity (m_g) and Gas Compressibility Factor also vary with pressure but not to the same degree. Pseudo-Pressure and pseudo-time (t_a) are used to deal with these changing properties and linearize the flow equations for gas. With the introduction of pseudo-pressure and pseudo-time, the gas flow equation can be written in a manner similar to the liquid equation. Therefore, the liquid flow solution can be used for gas well test analysis and forecasting provided pressure is replaced by pseudo-pressure, and time is replaced by pseudo-time as follows:

*Pseudo-Pressure = Constant * log (Pseudo-time) + ...*

Effect of Applying Pseudo-time: The effects of applying pseudo-time are significant for the following two conditions as illustrated in the subsequent plots:

- Low pressure at the wellbore
- Depleting reservoirs

The first plot shows that type curves can be inaccurate if pseudo-time is not used. The second plot shows that the buildup pressure does not reach up to the average reservoir pressure, when pseudo-time is not used. Thus, it is imperative that pseudo-pressure and pseudo-time be used for accurate analysis, modelling, and forecasting of gas production.

Development of Pseudo-time: It should be noted that the concept of pseudo-time is not amenable to a completely rigorous solution, as is the case for pseudo-pressure, because the gas properties change with pressure, not time. The form of the diffusivity equation dictates which properties are grouped with pressure and which are grouped with time.

Pseudo-time was developed by Agarwal (1980) and he defined pseudo-time in terms of the viscosity (μ_g) and compressibility (c_t) at the wellbore. This definition accounted for the large change in gas compressibility (c_g) that occurs at low pressures (early time in a buildup). It had little effect on late time data, and was generally used for buildups only. In the 90's, when the gas flow equations were being used for analyzing or forecasting data affected by reservoir depletion, it was realized that the Agarwal definition of pesudo-time for buildups, was inappropriate for boundary dominated flow (depleting systems). Moreover, the Agarwal pseudo-time definition did not solve the problem, because it was using a simplified version of the total system compressibility (c_t). Blasingame et al. introduced a new definition of pseudo-time to account for the depletion effects.

Instead of defining the pseudo-time transformation in terms of wellbore conditions like Agarwal did, they defined it in terms of the average reservoir pressure. This pseudo-time is best described as material balance pseudo-time, which is appropriate for boundary dominated flow. Fekete's Rahman, Mattar and Zaoral showed that a rigorous definition of compressibility, which honours material balance, is required because fluid saturations can change over time. Fekete's

Anderson and Mattar found that, in reservoir with significant transient flow, it was more appropriate to define pseudo-time in terms of the average pressure within the region of investigation rather than the average reservoir pressure.

Two different pseudo-time formulations are needed, one for buildup and a different one for drawdown. The equations look similar, but they are radically different from each other.

Drilling Engineering

Drilling engineering is a subset of petroleum engineering. Drilling engineers design and implement procedures to drill wells as safely and economically as possible. They work closely with the drilling contractor, service contractors, and compliance personnel, as well as with geologists and other technical specialists. The drilling engineer has the responsibility for ensuring that costs are minimized while getting information to evaluate the formations penetrated, protecting the health and safety of workers and other personnel, and protecting the environment.

Overview

The planning phases involved in drilling an oil or gas well typically involve estimating the value of sought reserves, estimating the costs to access reserves, acquiring property by a mineral lease, a geological survey, a well bore plan, and a layout of the type of equipment required to reach the depth of the well. Drilling engineers are in charge of the process of planning and drilling the wells. Their responsibilities include:

1. Designing well programmes (e.g., casing sizes and setting depths) to prevent blowouts (uncontrolled well-fluid release) while allowing adequate formation evaluation.
2. Designing or contributing to the design of casing strings and cementing plans, directional drilling plans, drilling fluids programmes, and drill string and drill bit programmes.
3. Specifying equipment, material and ratings and grades to be used in the drilling process.
4. Providing technical support and audit during the drilling process.
5. Performing cost estimates and analysis.
6. Developing contracts with vendors.

Drilling engineers are often degreed as petroleum engineers, although they may come from other technical disciplines (e.g., mechanical engineering or geology) and subsequently be trained by an oil and gas company. They also may have practical experience as a rig hand or mudlogger or mud engineer.

Measurement While Drilling

Figure: *Mud Moter & MWD Tool*

MWD stands for Measurement While Drilling in the oil & gas industry. The simplest way to describe MWD is to relate it to the measurements a pilot takes. A pilot needs to know the direction they are flying (North, South, East, or West), the angle they are flying at (up, down, or horizontal), and what type of skies they will be flying through (rough, choppy, cloudy, rainy, etc.). Like a pilot a directional driller needs to know these items about the ground formations that they are drilling through. MWD provides this information. Previous to MWD measurements were taken at various parts of the drilling process, but MWD has allowed these measurements to be sent to the surface continuously while the hole is being drilled. This allows for faster drilling, more accurate drilling, and safer drilling.

Beyond the basic concept MWD is a system developed to perform drilling related measurements downhole that are transmitted to the surface while drilling a well. MWD tools are installed as part of the bottom hole assembly (BHA) near the drill bit. The tools are either contained inside a thick walled, drill collar (drill collars are typically used to add weight for drilling) or they are built directly into the collars at a factory prior to arriving on the drilling location.

MWD systems can take several measurements such as Gamma Ray, compass direction (shown as azimuth), tool face (the direction that your bit is pointing), borehole pressure, temperature, vibration, shock, torque, etc. The MWD also provides the means of communication for operating rotary steering tools (RSTs).

The measured results are stored in MWD tools and some of the results can be transmitted digitally to surface using mud pulser telemetry through the mud or other advanced technology such as electromagnetic(EM) frequency communications or wired drill pipe.

Types of Information Transmitted

Directional Information: MWD tools are generally capable of taking directional surveys in real time. The tool uses accelerometers and magnetometers to measure the inclination and azimuth of the wellbore at that location, and they then transmit that information to the surface. With a series of surveys; measurements of inclination, azimuth, and tool face, at appropriate intervals (anywhere from every 30 ft (i.e. 10m) to every 500 ft), the location of the wellbore can be calculated.

By itself, this information allows operators to prove that their well does not cross into areas that they are not authorized to drill. However, due to the cost of MWD systems, they are not generally used on wells intended to be vertical. Instead, the wells are surveyed after drilling through the use of Multishot Surveying Tools lowered into the drillstring on slickline or wireline.

The primary use of real-time surveys is in Directional Drilling. For the Directional Driller to steer the well towards a target zone, he must know where the well is going, and what the effects of his steering efforts are.

MWD tools also generally provide toolface measurements to aid in directional drilling using downhole mud motors with bent subs or bent housings. For more information on the use of toolface measurements.

Drilling Mechanics Information

MWD tools can also provide information about the conditions at the drill bit. This may include:

- Rotational speed of the drillstring
- Smoothness of that rotation
- Type and severity of any vibration downhole
- Downhole temperature
- Torque and Weight on Bit, measured near the drill bit
- Mud flow volume

Figure: *Mud Moter & MWD*

Use of these information can allow the operator to drill the well more efficiently, and to ensure that the MWD tool and any other downhole tools, such as a Mud motor, Rotary Steerable Systems, and LWD tools, are operated within their technical specifications to prevent tool failure. This information also is valuable to Geologists responsible for the well information about the formation which is being drilled.

Formation Properties

Many MWD tools, either on their own, or in conjunction with separate Logging While Drilling tools, can take measurements of formation properties. At the surface, these measurements are assembled into a log, similar to one obtained by wireline logging.

LWD Logging While Drilling tools are able to measure a suite of geological characteristics including- density, porosity, resistivity, acoustic-caliper, inclination at the drill bit (NBI), magnetic resonance and formation pressure.

The MWD tool allows these measurements to be taken and evaluated while the well is being drilled. This makes it possible to perform Geosteering, or Directional Drilling based on measured formation properties, rather than simply drilling into a preset target.

Most MWD tools contain an internal Gamma Ray sensor to measure natural Gamma Ray values. This is because these sensors are compact, inexpensive, reliable, and can take measurements through unmodified drill collars. Other measurements often require separate Logging While Drilling tools, which communicate with the MWD tools downhole through internal wires.

Measurement while drilling can be cost-effective in exploration wells, particularly in areas of the Gulf of Mexico where wells are drilled in areas of salt diapirs. The resistivity log will detect penetration into salt, and early detection prevents salt damage to bentonite drilling mud.

Data Transmission Methods

Mud Pulse Telemetry: This is the most common method of data transmission used by MWD (Measurement While Drilling) tools. Downhole a valve is operated to restrict the flow of the drilling mud (slurry) according to the digital information to be transmitted. This creates pressure fluctuations representing the information. The pressure fluctuations propagate within the drilling fluid towards the surface where they are received from pressure sensors. On the surface, the received pressure signals are processed by computers to reconstruct the information. The technology is available in three varieties - *positive* pulse, *negative* pulse, and *continuous wave.*

Positive Pulse: Positive Pulse tools briefly close and open the valve to restrict the mud flow within the drill pipe. This produces an increase in pressure that can be seen at surface. Line codes are used to represent the digital information in form of pulses.

Negative Pulse: Negative pulse tools briefly open and close the valve to release mud from inside the drillpipe out to the annulus. This produces a decrease in pressure that can be seen at surface. Line codes are used to represent the digital information in form of pulses.

Continuous Wave: Continuous wave tools gradually close and open the valve to generate sinusoidal pressure fluctuations within the drilling fluid. Any digital modulation scheme with a continuous phase can be used to impose the information on a carrier signal. The most widely used modulation scheme is continuous phase modulation.

When underbalanced drilling is used, mud pulse telemetry can become unusable. This is because usually in order to reduce the equivalent density of the drilling mud a compressible gas is injected into the mud. This causes high signal attenuation which drastically reduces the ability of the mud to transmit pulsed data. In this case it is necessary to use methods different from mud pulse telemetry, such as electromagnetic waves propagating through the formation or wired drill pipe telemetry.

Current mud pulse telemetry technology offers a bandwidths of up to 40 bit/s. The data rate drops with increasing length of the wellbore and is typically as low as 1.5 bit/s - 3.0 bit/s. (bits per second) at a depth of 35,000 ft - 40,000 ft (10668 m - 12192 m).

Surface to down hole communication is typically done via changes to drilling parameters, i.e. change of the rotation speed of the drill string or change of the mud flow rate. Making changes to the drilling parameters in order to send information can require interruption of the drilling process, which is unfavourable due to the fact that it causes non-productive time.

Electromagnetic Telemetry (EM Tool)

These tools incorporate an electrical insulator in the drillstring. To transmit data, the tool generates an altered voltage difference between the top part (the main drillstring, above the insulator), and the bottom part (the drill bit, and other tools located below the insulator of the MWD tool). On surface a wire is attached to the wellhead, which makes contact with the drillpipe at the surface. A second wire is attached to a rod driven into the ground some distance away. The wellhead and the ground rod form the two electrodes of a dipole antenna. The voltage difference between the two electrodes is the receive signal that is decoded by a computer.

The EM tool generates voltage differences between the drillstring sections in the pattern of very low frequency (2–12 Hz) waves. The data is imposed on the waves through digital modulation.

This system generally offers data rates of up to 10 bits per second. In addition, many of these tools are also capable of receiving data from the surface in the same way, while mud pulse-based tools rely on changes in the drilling parameters, such as rotation speed of the drillstring or the mud flow rate, to send information from the surface to downhole tools. Making changes to the drilling parameters in order to send information to the tools generally interrupts the drilling process, causing lost time.

Compared to mud pulse telemetry, electronic pulse telemetry is more effective in certain specialized situations, such as underbalanced drilling or when using air as drilling fluid. However, it generally falls short when drilling exceptionally deep wells, and the signal can lose strength rapidly in certain types of formations, becoming undetectable at only a few thousand feet of depth.

Wired Drill Pipe

Several oilfield service companies are currently developing wired drill pipe systems. These systems use electrical wires built into every component of the drillstring, which carry electrical signals directly to the surface. These systems promise data transmission rates orders of magnitude greater than anything possible with mud pulse or electromagnetic telemetry, both from the downhole tool to the surface, and from the surface to the downhole tool. The IntelliServ wired pipe network, offering data rates upwards of 1 megabit per second, became commercial in 2006. Representatives from BP America, StatoilHydro, Baker Hughes INTEQ, and Schlumberger presented three success stories using this system, both onshore and offshore, at the March, 2008 SPE/IADC Drilling Conference in Orlando, Florida.

Retrievable Tools

MWD tools may be semi-permanently mounted in a drill collar (only removable at servicing facilities), or they may be self-contained and wireline retrievable. Retrievable tools, sometimes known as *Slim Tools*, can be retrieved and replaced using wireline through the drill string. This generally allows the tool to be replaced much faster in case of failure, and it allows the tool to be recovered if the drillstring becomes stuck. Retrievable tools must be much smaller, usually about 2 inches or less in diameter, though their length may be 20 feet or more. The small size is necessary for the tool to fit through the drillstring, however, it also limits the tool's capabilities. For example, slim tools are not capable of sending data at the same rates as collar mounted tools, and they are also more limited in their ability to communicate with and supply electrical power to other LWD tools.

Collar-mounted tools, also known as *Fat Tools*, cannot generally be removed from their drill collar at the wellsite. If the tool fails, the entire drillstring must be pulled out of the hole to replace it. However, without the need to fit through the drillstring, the tool can be larger and more capable.

The ability to retrieve the tool via wireline is often useful. For example, if the drillstring becomes stuck in the hole, then retrieving the tool via wireline will save a substantial amount of money compared to leaving it in the hole with the stuck portion of the drillstring. However, there are some limitations on the process.

Limitations

Retrieving a tool using wireline is not necessarily faster than pulling the tool out of the hole. For example, if the tool fails at 1,500 ft (460 m) while drilling with a triple rig (able to trip 3 joints of pipe, or about 90 ft (30 m) feet, at a time), then it would generally be faster to pull the tool out of the hole than it would be to rig up wireline and retrieve the tool, especially if the wireline unit must be transported to the rig.

Wireline retrievals also introduce additional risk. If the tool becomes detached from the wireline, then it will fall back down the drillstring. This will generally cause severe damage to the tool and the drillstring components in which it seats, and will require the drillstring to be pulled out of the hole to replace the failed components, thus resulting in a greater total cost than pulling out of the hole in the first place. The wireline gear might also fail to latch onto the tool, or in the case of a severe failure, might bring only a portion of the tool to the surface. This would require the drillstring to be pulled out of the hole to replace the failed components, thus making the wireline operation a waste of time.

Geosteering

In the process of drilling a borehole, geosteering is the act of adjusting the borehole position (inclination and azimuth angles) *on the fly* to reach one or more geological targets. These changes are based on geological information gathered while drilling.

Description

From 2D and 3D models of underground substructures, deviated wells (2D and 3D) are planned in advance to achieve specific goals: exploration, fluids production, fluids injection or technical.

A *well plan* is a continuous succession of straight and curved lines representing the geometrical figure of the expected well path. A well plan is always projected on vertical and horizontal maps.

While the borehole is being drilled according to the well plan, new geological information is gathered from mud logging, measurement while drilling (MWD) and logging while drilling (LWD). These usually show some differences from what is expected from the model. As the model is continuously updated with the new geological information (formation evaluation) and the borehole position (*well deviation survey*), changes start to appear in the geological substructures and can lead to the well plan being updated to reach the corrected geological targets.

Logging while Drilling

Logging While Drilling (LWD) is a technique of conveying well logging tools into the well borehole downhole as part of the bottom hole assembly (BHA).

LWD tools work with its Measurement While Drilling (MWD) system to transmit partial or complete measurement results to the surface via typically a drilling mud pulser or other improved techniques, while LWD tools are still in the borehole, which is called "Real Time Data". Complete measurement results can be downloaded from LWD tools after they are pulled out of hole, which is called "Memory Data".

LWD technology was developed originally as an enhancement to the earlier MWD technology to completely or partially replace wireline logging operation. With the improvement of the technology in the past decades, LWD is now widely used for drilling (including geosteering), and formation evaluation (especially for real time and high angle wells).

Available LWD Measurements

LWD technology was originally developed to partially or completely replace wireline logging. Over the years, more of the measurements have been made available in LWD. Certain new measurements are also development in LWD only. The following is an incomplete list of available measurement in LWD technology.

- Natural Gamma Ray (GR)
  - Total Gamma Ray
  - Spectral Gamma Ray
  - Azimuthal Gamma Ray
  - Gamma ray close to drill bit.
- Density and Photoelectric Index
- Neutron Porosity
- Borehole Caliper
  - Ultra sonic azimuthal caliper.
  - Density Caliper
- Resistivity (ohm-m)
  - Attenuation and phase shift resistivities at different transmitter spacings and frequencies.
  - Resistivity at the drill bit.
  - Deep directional resistivities.
- Sonic
  - Compressional Slowness(Δtc)
  - Shear Slowness (Δts)
- Borehole Images
  - Density Borehole Image
  - Resistivity Borehole Image
- Formation Tester and Sampler
  - Formation Pressure
  - Formation Fluid Sample
- Nuclear Magnetic Resonance (NMR)
- Seismic While Drilling (SWD)
  - Drillbit-SWD

Chapter 3
Oil Well Testing Methods

Oil Well

An oil well is a general term for any boring through the Earth's surface that is designed to find and acquire petroleum oil hydrocarbons. Usually some natural gas is produced along with the oil. A well that is designed to produce mainly or only gas may be termed a gas well.

Usually attached to the oil derrick is the Oil well Dog House where the decisions are made for actions taken on the rig floor

Newer oil wells have hydraulic feet that are called Walking Rigs that allow for the rig to move several yards on its own without taking the derrick apart. This saves time as dual laterals can be drilled.

Figure: *The pumpjack, such as this one located south of Midland, Texas, is a common sight in West Texas*

History

The earliest known oil wells were drilled in China in 347 CE. These wells had depths of up to about 800 feet (240 m) and were drilled using bits attached to bamboo poles. The oil was burned to evaporate brine and produce salt. By the 10th century, extensive bamboo pipelines connected oil wells with salt springs. The ancient records of China and Japan are said to contain many allusions to the use of natural gas for lighting and heating. Petroleum was known as *burning water* in Japan in the 7th century.

According to Kasem Ajram, petroleum was distilled by the Persian alchemist Muhammad ibn Zakarîya Râzi (Rhazes) in the 9th century, producing chemicals such as kerosene in the alembic (*al-ambiq*), and which was mainly used for kerosene lamps. Arab and Persian chemists also distilled crude oil in order to produce flammable products for military purposes. Through Islamic Spain, distillation became available in Western Europe by the 12th century.

Some sources claim that from the 9th century, oil fields were exploited in the area around modern Baku, Azerbaijan, to produce naphtha for the petroleum industry. These fields were described by Marco Polo in the 13th century, who described the output of those oil wells as hundreds of shiploads. When Marco Polo in 1264 visited the Azerbaijani city of Baku, on the shores of the Caspian Sea, he saw oil being collected from seeps. He wrote that "on the confines toward Geirgine there is a fountain from which oil springs in great abundance, in as much as a hundred shiploads might be taken from it at one time."

Shallow pits were dug at the Baku seeps in ancient times to facilitate collecting oil, and hand-dug holes up to 35 metres (115 ft) deep were in use by 1594. These holes were essentially oil wells. Apparently 116 of these wells in 1830 produced 3,840 metric tons (about 28,000 barrels) of oil. Also, offshore drilling started up at Baku (then Russian Empire) at Bibi-Eibat field in 1846. In the New World, the first commercial oil well entered operation in Oil Springs, Ontario in 1858, while the first offshore oil well was drilled in 1896 at the Summerland Oil Field on the California Coast.

The earliest oil wells in modern times were drilled percussively, by hammering a cable tool into the earth. Soon after, cable tools were replaced with rotary drilling, which could drill boreholes to much greater depths and in less time. The record-depth Kola Borehole used non-rotary mud motor drilling to achieve a depth of over 12,000 metres (39,000 ft).

Until the 1970s, most oil wells were vertical, although lithological and mechanical imperfections cause most wells to deviate at least slightly from true vertical.

However, modern directional drilling technologies allow for strongly deviated wells which can, given sufficient depth and with the proper tools, actually become horizontal. This is of great value as the reservoir rocks which contain hydrocarbons are usually horizontal, or sub-horizontal; a horizontal wellbore placed in a production zone has more surface area in the production zone than a vertical well, resulting in a higher production rate.

The use of deviated and horizontal drilling has also made it possible to reach reservoirs several kilometers or miles away from the drilling location (extended reach drilling), allowing for the production of hydrocarbons located below locations that are either difficult to place a drilling rig on, environmentally sensitive, or populated.

Life of a Well

The creation and life of a well can be divided up into five segments:

- Planning
- Drilling
- Completion
- Production
- Abandonment

Drilling

The well is created by drilling a hole 5 to 50 inches (127.0 mm to 914.4 mm) in diameter into the earth with a drilling rig that rotates a drill string with a bit attached. After the hole is drilled, sections of steel pipe (casing), slightly smaller in diameter than the borehole, are placed in the hole. Concrete may be placed between the outside of the casing and the borehole. The casing provides structural integrity to the newly drilled wellbore, in addition to isolating potentially dangerous high pressure zones from each other and from the surface.

With these zones safely isolated and the formation protected by the casing, the well can be drilled deeper (into potentially more-unstable and violent formations) with a smaller bit, and also cased with a smaller size casing. Modern wells often have two to five sets of subsequently smaller hole sizes drilled inside one another, each cemented with casing.

To Drill the Well

Figure: *Mud log in process, a common way to study the lithology when drilling oil wells*

- The drill bit, aided by the weight of thick walled pipes called "drill collars" above it, cuts into the rock. There are different types of drill bit; some cause the rock to disintegrate by compressive failure, while others shear slices off the rock as the bit turns.
- Drilling fluid, a.k.a. "mud", is pumped down the inside of the drill pipe and exits at the drill bit. Drilling mud is a complex mixture of fluids, solids and chemicals that must be carefully tailored to provide the correct physical and chemical characteristics required to safely drill the well. Particular functions of the drilling mud include cooling the bit, lifting rock cuttings to the surface, preventing destabilisation of the rock in the wellbore walls and overcoming the pressure of fluids inside the rock so that these fluids do not enter the wellbore.
- The generated rock "cuttings" are swept up by the drilling fluid as it circulates back to surface outside the drill pipe. The fluid then goes through "shakers" which strain the cuttings from the good fluid which is returned to the pit. Watching for abnormalities in the returning cuttings and monitoring pit volume or rate of returning fluid are imperative to catch "kicks"

early. A "kick" is when the formation pressure at the depth of the bit is more than the hydrostatic head of the mud above, which if not controlled temporarily by closing the blowout preventers and ultimately by increasing the density of the drilling fluid would allow formation fluids and mud to come up through the annulus uncontrollably.

- The pipe or drill string to which the bit is attached is gradually lengthened as the well gets deeper by screwing in additional 30-foot (9 m) sections or "joints" of pipe under the kelly or topdrive at the surface. This process is called making a connection, or "tripping". Joints can be combined for more efficient tripping when pulling out of the whole by creating stands of multiple joints. A conventional triple, for example, would pull pipe out of the hole three joints at a time and stack them in the derrick. Many modern rigs, called "super singles," trip pipe one at a time, laying it out on racks as they go.

This process is all facilitated by a drilling rig which contains all necessary equipment to circulate the drilling fluid, hoist and turn the pipe, control downhole, remove cuttings from the drilling fluid, and generate on-site power for these operations.

Completion (Oil and Gas Wells)

Figure: *Perforated Shoe*

In petroleum production, completion is the process of making a well ready for production (or injection). This principally involves preparing the bottom of the hole to the required specifications, running in the production tubing and its associated down hole tools as well as perforating and stimulating as required. Sometimes, the process of running in and cementing the casing is also included.

Lower Completion

This refers to the portion of the well across the production or injection zone. The well designer has many tools and options available to design the lower completion according to the conditions of the reservoir. Typically, the lower completion is set across the productive zone using a liner hanger system, which anchors the lower completion to the production casing string. The broad categories of lower completion are listed below.

Barefoot Completion

This type is the most basic, but can be a good choice for hard rock, multi-laterals and underbalance drilling. It involves leaving the productive reservoir section without any tubulars. This effectively removes control of flow of fluids from the formation; it is not suitable for weaker formations which might require sand control, nor for formations requiring selective isolation of oil, gas and water intervals. However, advances in interventions such as coiled tubing and tractors means that barefoot wells can be successfully produced.

Open Hole

The production casing is set above the zone of interest before drilling the zone. The zone is open to the well bore. In this case little expense is generated with perforations log interpretation is not critical. The well can be deepened easily and it is easily converted to screen and liner. However, excessive gas and water production is difficult to control, and may require frequent clean outs. Also the interval can't be selectively stimulated.

Open Hole Completion

This designation refers to a range of completions where no casing or liner is cemented in place across the production zone. In competent formations the zone might be left entirely bare, but usually some sort of sand-control and/or flow-control means are incorporated. Openhole completions have seen significant uptake in recent years, and there are many configurations, often developed to address specific reservoir challenges. There have been many recent developments that have boosted the success of openhole completions, and they also tend to be popular in horizontal wells, where cemented installations are more expensive and technically more difficult. The common options for openhole completions are:

Pre-holed Liner

Also often called pre-drilled liner. The liner is prepared with multiple small drilled holes, then set across the production zone to provide wellbore stability and an intervention conduit. Pre-holed liner is often combined with openhole packers, such as swelling elastomers, mechanical packers or external casing packers, to provide zonal segregation and isolation. It is now quite common to see a combination of pre-holed liner, solid liner and swelling elastomer packers to provide an initial isolation of unwanted water or gas zones. Multiple sliding sleeves can also be used in conjunction with openhole packers to provide considerable flexibility in zonal flow control for the life of the wellbore.

This type of completion is also being adopted in some water injection wells, although these require a much greater performance envelope for openhole packers, due to the considerable pressure and temperature changes that occur in water injectors.

Openhole completions (in comparison with cemented pipe) require better understanding of formation damage, wellbore clean-up and fluid loss control. A key difference is that perforating penetrates through the first 6-18 inches (15–45 cm) of formation around the wellbore, whilst openhole completions require the reservoir fluids to flow through all of the filtrate-invaded zone around the wellbore and lift-off of the mud filter cake.

Many openhole completions will incorporate fluid loss valves at the top of the liner to provide well control whilst the upper completion is run. There are an increasing number of ideas coming into the market place to extend the options for openhole completions; for example, electronics can be used to actuate a self-opening or self-closing liner valve. This might be used in an openhole completion to improve clean-up, by bringing the well onto production from the toe-end for 100 days, then self-opening the heel-end. Inflow control devices and intelligent completions are also installed as openhole completions.

Pre-holed liner may provide some basic control of solids production, where the wellbore is thought to fail in aggregated chunks of rubble, but it is not typically regarded as a sand control completion.

Slotted Liner

Slotted liners can be selected as an alternative to pre-holed liner, sometimes as a personal preference or from established practice on a field. It can also be selected to provide a low cost control of sand/solids production. The slotted liner is machined with multiple longitudinal slots, for example 2 mm x 50mm, spread across the length and circumference of each joint. Recent advances in laser cutting means that slotting can now be done much cheaper to much smaller slot widths and in some situation slotted liner is now used for the same functionality as sand control screens.

Openhole Sand Control

This is selected where the liner is required to mechanically hold-back the movement of formation sand. There are many variants of openhole sand control, the three popular choices being stand-alone screens, openhole gravel packs (also known as external gravel packs, where a sized sand 'gravel' is placed as an annulus around the sand control screen) and expandable screens. Screen designs are mainly wire-wrap or premium; wire-wrap screens use spiral-welded corrosion-resistant wire wrapped around a drilled basepipe to provide a consistent small helical gap (such as 0.012-inch (0.30 mm), termed 12 gauge). Premium screens use a woven metal cloth wrapped around a basepipe. Expandable screens are run to depth before being mechanically swaged to a larger diameter. Ideally, expandable screens will be swaged until they contact the wellbore wall.

Horizontal open hole completions

This is the most common open hole completion used today. It is basically the same described on the vertical open hole completion but on a horizontal well it enlarges significantly the contact with the reservoir, increasing the production or injection rates of your well. Sand control on a horizontal well is completely different from a vertical well. We can no long rely on the gravity for the gravel placement. Most service companies uses an alpha and beta wave design to cover the total length of the horizontal well with gravel. It's known that very long wells (around 6000 ft) were successfully gravel packed in many occasions, including deepwater reservoirs in Brazil.

Liner Completions

In this case the casing is set above the primary zone. An un-cemented screen and liner assembly is installed across the pay section. This technique minimizes formation damage and gives the ability to control sand. It also makes cleanout easy. Perforating expense is also low to non-existent. However gas and water build up is difficult to control and selective stimulation not possible the well can't be easily deepened and additional rig time may be needed.

Perforated Liner

Casing is set above the producing zone, the zone is drilled and the liner casing is cemented in place. The liner is then perforated for production. This time additional expense in perforating the casing is incurred, also log interpretation is critical and it may be difficult to obtain good quality cement jobs.

Perforated Casing

Production casing is cemented through the zone and the pay section is selectively perforated. Gas and water are easily controlled as is sand. The formation can be selectively stimulated and the well can be deepened. This selection is adaptable to other completion configurations and logs are available to assist casing decisions. Much better primary casing. It can however cause damage to zones and needs good log interpretation. The perforating cost can be very high.

Cased Hole Completion

This involves running casing or a liner down through the production zone, and cementing it in place. Connection between the well bore and the formation is made by perforating. Because perforation intervals can be precisely positioned, this type of completion affords good control of fluid flow, although it relies on the quality of the cement to prevent fluid flow behind the liner. As such it is the most common form of completion...

Conventional Completions

- *Casing flow* means that the producing fluid flow has only one path to the surface through the casing.
- *Casing and tubing flow* means that the there is tubing within the casing that allows fluid to reach the surface. This tubing can be used as a kill string for chemical injection. The tubing may have a "no-go" nipple at the end as a means of pressure testing.
- *Pumping flow* the tubing and pump are run to a depth beneath the working fluid. The pump and rod string are installed concentrically within the tubing. A tubing anchor prevents tubing movement while pumping.

- *Tubing flow* a tubing string and a production packer are installed. The packer means that all the flow goes through the tubing. Within the tubing you can mount a combination of tools that will help to control fluid flow through the tubing.
- *Gas lift well* gas is fed into valves installed in mandrels in the tubing strip. The hydrostatic head is lowered and the fluid is gas lifted to the surface.
- *Single-well alternate completions* in this instance there is a well with two zones. In order to produce from both the zones are isolated with packers. Blast joints may be used on the tubing within the region of the perforations. These are thick walled subs that can withstand the fluid abrasion from the producing zone. This arrangement can also work if you have to produce from a higher zone given the depletion of a lower zone. The tubing may also have flow control mechanism.
- *Single-well concentric kill string* within the well a small diameter concentric kill string is used to circulate kill fluids when needed.
- *Single-well 2-tubing completion* in this instance 2 tubing strings are inserted down 1 well. They are connected at the lower end by a circulating head. Chemicals can be circulated down one tube and production can continue up the other.

Completion Components

The upper completion refers to all components from the bottom of the production tubing upwards. Proper design of this "completion string" is essential to ensure the well can flow properly given the reservoir conditions and to permit any operations as are deemed necessary for enhancing production and safety.

Wellhead

A wellhead is the component at the surface of an oil or gas well that provides the structural and pressure-containing interface for the drilling and production equipment.

The primary purpose of a wellhead is to provide the suspension point and pressure seals for the casing strings that run from the bottom of the hole sections to the surface pressure control equipment. While drilling the oil well, surface pressure control is provided by a blowout preventer (BOP). If the pressure is not contained during drilling operations by the column of drilling fluid, casings, wellhead, and BOP, a well blowout could occur. Once the well has been drilled, it is *completed* to provide an interface with the reservoir rock and a tubular conduit for the well fluids.

The surface pressure control is provided by a Christmas tree, which is installed on top of the wellhead, with isolation valves and choke equipment to control the flow of well fluids during production.

Figure: *Wellhead gas storage, Etzel Germany*

Wellheads are typically welded onto the first string of casing, which has been cemented in place during drilling operations, to form an integral structure of the well.

In exploration wells that are later abandoned, the wellhead may be recovered for refurbishment and re-use. Offshore, where a wellhead is located on the production platform it is called a *surface wellhead*, and if located beneath the water then it is referred to as a subsea wellhead or mudline wellhead.

Functions

A wellhead serves numerous functions, some of which are:

1. Provide a means of casing suspension. (Casing is the permanently installed pipe used to line the well hole for pressure containment and collapse prevention during the drilling phase).

2. Provides a means of tubing suspension. (Tubing is removable pipe installed in the well through which well fluids pass).
3. Provides a means of pressure sealing and isolation between casing at surface when many casing strings are used.
4. Provides pressure monitoring and pumping access to annuli between the different casing/tubing strings.
5. Provides a means of attaching a blowout preventer during drilling.
6. Provides a means of attaching a Christmas tree for production operations.
7. Provides a reliable means of well access.
8. Provides a means of attaching a well pump,

Components

The primary components of a wellhead system are:

- casing head
- casing spools
- casing hangers
- packoffs (isolation) seals
- bowl protectors / wear bushings
- test plugs
- mudline suspension systems
- tubing heads
- tubing hangers
- tubing head adapters.

Design Specification

The oil industry specifications for wellhead systems (materials, dimensions, test procedures and pressure ratings etc.) are :

1. API 6A Specification for Wellhead and Christmas Tree Equipment
2. ISO 10423 Wellhead and Christmas Tree Equipment

In general well heads are 5 nominal ratings of wellheads 2,3,5,10 and 15 (x1000)psi working pressure. They have an operating temperature range of -50 to +250 degrees Fahrenheit. They are used in conjunction with ring type seal gaskets.

In general the yield strength of the materials range from 36000 to 75000psi.

Christmas Tree (Oil Well)

Figure: *Oil well Christmas tree*

In petroleum and natural gas extraction, a Christmas tree, or "tree" (not "wellhead" as sometimes incorrectly referred to), is an assembly of valves, spools, and fittings used for an oil well, gas well, water injection well, water disposal well, gas injection well, condensate well and other types of wells. It was named for its crude resemblance to a decorated tree.

Overview

Christmas trees are used on both surface and subsea wells. It is common to identify the type of tree as either "subsea tree" or "surface tree". Each of these classifications has a number of variations. Examples of subsea include conventional, dual bore, mono bore, TFL (through flow line), horizontal, mudline, mudline horizontal, side valve, and TBT (through-bore tree) trees. The deepest installed subsea tree is in the Gulf of Mexico at approximately 9,000 feet (2,700 m). (Current technical limits are up to around 3000 metres and working temperatures of -50°F to 350°F with a pressure of up to 15,000 psi.)

The primary function of a tree is to control the flow, usually oil or gas, out of the well. (A tree may also be used to control the injection of gas or water into a non-producing well in order to enhance production rates of oil from other wells.) When the well and facilities are ready to produce and receive oil or gas, tree valves are opened and the formation fluids are allowed to go through a flow line. This leads to a processing facility, storage depot and/or other pipeline eventually leading to a refinery or distribution centre (for gas). Flow lines on subsea wells usually lead to a fixed or floating production platform or to a storage ship or barge, known as a floating storage offloading vessel (FSO), or floating processing unit (FPU), or floating production, storage and offloading vessel(FPSO).

A tree often provides numerous additional functions including chemical injection points, well intervention means, pressure relief means, monitoring points (such as pressure, temperature, corrosion, erosion, sand detection, flow rate, flow composition, valve and choke position feedback), and connection points for devices such as down hole pressure and temperature transducers (DHPT). On producing wells, chemicals or alcohols or oil distillates may be injected to preclude production problems (such as blockages).

Functionality may be extended further by using the control system on a subsea tree to monitor, measure, and react to sensor outputs on the tree or even down the well bore. The control system attached to the tree controls the downhole safety valve (SCSSV, DHSV, SSSV) while the tree acts as an attachment and conduit means of the control system to the downhole safety valve. Tree complexity has increased over the last few decades.

They are frequently manufactured from blocks of steel containing multiple valves rather than being assembled from individual flanged components. This is especially true in subsea applications where the resemblance to Christmas trees no longer exists given the frame and support systems into which the main valve block is integrated.

Note that a tree and wellhead are separate pieces of equipment not to be mistaken as the same piece. The Christmas tree is installed on top of the wellhead.

A wellhead is used without a Christmas tree during drilling operations, and also for riser tie-back situations that later would have a tree installed at riser top. Wells being produced with rod pumps (pump jacks, nodding donkeys, and so on) frequently do not utilize any tree owing to no pressure-containment requirement.

Valves

Subsea and surface trees have a large variety of valve configurations and combinations of manual and/or actuated (hydraulic or pneumatic) valves. Examples are identified in API Specifications 6A and 17D. A basic surface tree consists of two or three manual valves (usually gate valves because of their flow characteristics, i.e. low restriction to the flow of fluid when fully open).

A typical sophisticated surface tree will have at least four or five valves, normally arranged in a crucifix type pattern (hence the endurance of the term "Christmas tree"). The two lower valves are called the master valves (upper and lower respectively). Master valves are normally in the fully open position and are *never* opened or closed when the well is flowing (except in an emergency) to prevent erosion of the valve sealing surfaces.

The lower master valve will normally be manually operated, while the upper master valve is often hydraulically actuated, allowing it to be used as a means of remotely shutting in the well in the event of emergency. An actuated wing valve is normally used to shut in the well when flowing, thus preserving the master valves for positive shut off for maintenance purposes. Hydraulic operated wing valves are usually built to be fail safe closed, meaning they require active hydraulic pressure to stay open. This feature means that if control fluid fails the well will automatically shut itself in without operator action.

The right hand valve is often called the flow wing valve or the production wing valve, because it is in the flowpath the hydrocarbons take to production facilities (or the path water or gas will take from production to the well in the case of injection wells). The left hand valve is often called the kill wing valve. It is primarily used for injection of fluids such as corrosion inhibitors or methanol to prevent hydrate formation. In the North Sea, it is called the non-active side arm (NASA). It is typically manually operated.

The valve at the top is called the swab valve and lies in the path used for well interventions like wireline and coiled tubing. For such operations, a lubricator is rigged up onto the top of the tree and the wire or coil is lowered through the lubricator, past the swab valve and into the well. This valve is typically manually operated. Some trees have a second swab valve, the two arranged one on top of the other. The intention is to allow rigging down equipment from the top of the tree with the well flowing while still preserving the Two-barrier rule.

With only a single swab valve, the upper master valve is usually closed to act as the second barrier, forcing the well to be shut in for a day during rig down operations. However, avoiding delaying production for a day is usually too small a gain to be worth the extra expense of a having a Christmas tree with a second swab valve.

Subsea trees are available in either vertical or horizontal configurations with further speciality available such as dual bore, monobore, concentric, drill-through, mudline, guidlineless or guideline. Subsea trees may range in size and weight from a few tons to approximately 70 tons for high pressure, deepwater (>3000 feet) guidelineless applications. Subsea trees contain many additional valves and accessories compared to Surface trees. Typically a subsea tree would have a choke (permits control of flow), a flowline connection interface (hub, flange or other connection), subsea control interface (direct hydraulic, electro hydraulic, or electric) and sensors for gathering data such as pressure, temperature, sand flow, erosion, multiPhase flow, single phase flow such as water or gas.

Tubing Hanger

A tubing hanger is a component used in the completion of oil and gas production wells. It is set in the tree or the wellhead and suspends the production tubing and/or casing. Sometimes it provides porting to allow the communication of hydraulic, electric and other downhole functions, as well as chemical injection. It also serves to seal-in the annulus and production areas.

Production Tubing

Production tubing is a tube used in a wellbore through which production fluids are produced (travel).

Production tubing is run into the drilled well after the casing is run and cemented in place. Production tubing protects wellbore casing from wear, tear, corrosion, and deposition of by-products, such as sand / silt, paraffins, and asphaltenes. Along with other components that constitute the production string, it provides a continuous bore from the production zone to the wellhead through which oil and gas can be produced. It is usually between five and ten centimetres in diameter and is held inside the casing through the use of expandable packing devices. Purpose and design of production tubing is to enable quick, efficient, and safe installation, removal and re-installation. If there is more than one zone of production in the well, up to four lines of production tubing can be run.

Downhole Safety Valve

A downhole safety valve refers to a component on an oil and gas well, which acts as a failsafe to prevent the uncontrolled release of reservoir fluids in the event of a worst case scenario surface disaster. It is almost always installed as a vital component on the completion.

Operation

These valves are commonly uni-directional flapper valves which open downwards such that the flow of wellbore fluids tries to push it shut, while pressure from the surface pushes it open. This means that when closed, it will isolate the reservoir fluids from the surface. Most downhole safety valves are controlled hydraulically from the surface, meaning they are opened using a hydraulic connection linked directly to a well control panel.

When hydraulic pressure is applied down a control line, the hydraulic pressure forces a sleeve within the valve to slide downwards. This movement compresses a large spring and pushes the flapper downwards to open the valve. When hydraulic pressure is removed, the spring pushes the sleeve back up and causes the flapper to shut. In this way, it is fail-safe and will isolate the wellbore in the event of a loss of the wellhead. The full designation for a typical valve is 'tubing retrievable, surface controlled, subsurface safety valve', abbreviated to TR-SCSSV.

Positioning

The location of the downhole safety valve within the completion is a precisely determined parameter intended to optimise safety. There are arguments against it either being too high or too low in the well and so the final depth is a compromise of all factors. MMS regulations state that the valve must be placed no less than 100' below the mudline.

Reasons to Keep it High

The further down the well the DHSV is located, the greater the potential inventory of hydrocarbons above it when closed. This means that in the event of loss of containment at surface, there is more fluid to be spilled causing environmental damage, or in the worst case, more fuel for a fire. Therefore, placing the valve higher limits this hazard. Another reason relates to the hydraulic control line. Hydraulic pressure is required to keep the valve open as part of the fail-safe design. However, if the valve is too far down the well, then the portion of the control line below the surface is larger. If it is too large, then the weight of the hydraulic fluid alone may apply sufficient pressure to keep the valve open, even with the loss of surface pressurisation.

Reasons to Keep it Low

As part of the role of the DHSV to isolate the surface from wellbore fluids, it is necessary for the valve to be positioned away from the well where it could potentially come to harm. This implies that it must be placed subsurface in all circumstances, i.e. in offshore wells, not above the seabed. There is also the risk of cratering in the event of a catastrophic loss of the topside facility. The valve is specifically placed below the maximum depth where cratering is expected to be a risk. If there is a risk of methane hydrate (clathrate) plugs forming as the pressure changes through the valve due to Joule-Thomson cooling, then this is a reason to keep it low, where the rock is warmer than 30 C.

Deploying and Retrieving

Most downhole safety valves installed as part of the completion design are classed as "tubing retrievable". This means that they are installed as a component of the completion string and run in during completion. Retrieving the valve, should it malfunction, requires a workover. The full name for this most common type of downhole safety valve is a Tubing Retrievable Surface Controlled Sub-Surface Valve, shortened in completion diagrams to TRSCSSV. If a tubing retrievable valve fails, rather than go to the expense of a workover, a "wireline retrievable" valve may be used instead. This type of valve can fit inside the production tubing and is deployed on wireline after the old valve has been straddled open.

Legal Requirement

The importance of DHSVs is undisputed. Graphic images of oil wells in Kuwait on fire after the First Gulf War after their wellheads were removed, demonstrate the perils of not using the components (at the time, they were deemed unnecessary because they were onshore wells). It is, however, not a direct legal requirement in many places. In the United Kingdom, no law mandates the use of DHSVs. However, the 1974 Health & Safety at Work Act requires that measures are taken to ensure that the uncontrolled release of wellbore fluids is prevented even in the worst case.

The brilliance of the act is that it does not issue prescriptive guideline for how to achieve the goal of health and safety, but merely sets out the requirement that the goal be achieved. It is up to the oil companies to decide how to achieve it and DHSVs are an important component of that decision. As such, although not a legal requirement, it is company policy for many operators in the UKCS.

Issues

While the DHSV isolates the production tubing, a loss of integrity could allow wellbore fluid to bypass the valve and escape to surface through the annulus. For wells using gas lift, it is often a requirement to install a safety valve in the 'A' annulus of the well to ensure that the surface is protected from a loss of annulus containment. However, these valves are not as common and they are not necessarily installed at the same position in the well, meaning it is possible that fluids could snake their way around the valves to surface.

Annular Safety Valve

On wells with gas lift capability, many operators consider it prudent to install a valve, which will isolate the *A* annulus for the same reasons a DHSV may be needed to isolate the production tubing in order to prevent the inventory of natural gas downhole from becoming a hazard as it became on Piper Alpha.

Side Pocket Mandrel

This is a welded/machined product which contains a "side pocket" alongside the main tubular conduit. The side pocket, typically 1" or 1½" diameter is designed to contain gas lift valve, which allows hydrocarbon gas from the *A* annulus to be injected into the flow stream.

Submersible Pump

Figure: *One style of submersible pump for industrial use. Outlet pipe and electrical cable not connected.*

A submersible pump (or sub pump, electric submersible pump (ESP)) is a device which has a hermetically sealed motor close-coupled to the pump body. The whole assembly is submerged in the fluid to be pumped. The main advantage of this type of pump is that it prevents pump cavitation, a problem associated with a high elevation difference between pump and the fluid surface. Submersible pumps push fluid to the surface as opposed to jet pumps having to pull fluids. Submersibles are more efficient than jet pumps.

Working Principle

The submersible pumps used in ESP installations are multistage centrifugal pumps operating in a vertical position. Although their constructional and operational features underwent a continuous evolution over the years, their basic operational principle remained the same. Produced liquids, after being subjected to great centrifugal forces caused by the high rotational speed of the impeller, lose their kinetic energy in the diffuser where a conversion of kinetic to pressure energy takes place. This is the main operational mechanism of radial and mixed flow pumps.

The pump shaft is connected to the gas separator or the protector by a mechanical coupling at the bottom of the pump. Well fluids enter the pump through an intake screen and are lifted by the pump stages. Other parts include the radial bearings (bushings) distributed along the length of the shaft providing radial support to the pump shaft turning at high rotational speeds. An optional thrust bearing takes up part of the axial forces arising in the pump but most of those forces are absorbed by the protector's thrust bearing...

Applications

Submersible pumps are found in many applications. Single stage pumps are used for drainage, sewage pumping, general industrial pumping and slurry pumping. They are also popular with pond filters. Multiple stage submersible pumps are typically lowered down a borehole and used for water abstraction, water wells and in oil wells.

Submersible pumps are used in applications, including sewage treatment plants, seawater handling, groundwater pumping, fire fighting (since it is flame retardant cable), borehole drilling and pumping, water well and deep well drilling, offshore drilling rigs, artificial lifts, mine dewatering, irrigation systems, and water supply systems. Special attention to the type of ESP is required when using certain types of liquids. Pumps used for combustible liquids or for water that may be contaminated with combustible liquids must be designed not to ignite the liquid or vapors.

Use in Oil Wells

Submersible pumps are used in oil production to provide a relatively efficient form of "artificial lift", able to operate across a broad range of flow rates and depths. By decreasing the pressure at the bottom of the well (by lowering bottomhole flowing pressure, or increasing drawdown), significantly more oil can be produced from the well when compared with natural production. The pumps are typically electrically powered and referred to as Electrical Submersible Pumps (ESP). ESP systems consist of both surface components (housed in the production facility, for example an oil platform) and sub-surface components (found in the well hole). Surface components include the motor controller (often a variable speed controller), surface cables and transformers. Subsurface components typically include the pump, motor, seal and cables. A gas separator is sometimes installed.

The pump itself is a multi-stage unit with the number of stages being determined by the operating requirements. Each stage consists of a driven impeller and a diffuser which directs flow to the next stage of the pump. Pumps come in diameters from 90mm (3.5 inches) to 254mm (10 inches) and vary between 1 metre (3 ft) and 8.7 metres (29 ft) in length. The motor used to drive the pump is typically a three phase, squirrel cage induction motor, with a nameplate power rating in the range 7.5 kW to 560 kW (at 60 Hz).

New varieties of ESP can include a water/oil separator which permits the water to be reinjected into the reservoir without the need to lift it to the surface. There are at least 15 brands of oilfield esps used throughout the world. Until recently, ESPs had been highly costly to install due to the requirement of an electric cable downhole. This cable had to be wrapped around jointed tubing and connected at each joint. New coiled tubing umbilicals allow for both the piping and electric cable to be deployed with a single conventional coiled tubing unit.

The ESP system consists of a number of components that turn a staged series of centrifugal pumps to increase the pressure of the well fluid and push it to the surface. The energy to turn the pump comes from a high-voltage (3 to 5 kV) alternating-current source to drive a special motor that can work at high temperatures of up to 300 °F (149 °C) and high pressures of up to 5,000 psi (34 MPa), from deep wells of up to 12,000 feet (3.7 km) deep with high energy requirements of up to about 1000 horsepower (750 kW). ESPs have dramatically lower efficiencies with significant fractions of gas, greater than about 10% volume at the pump intake. Given their high rotational speed of up to 4000 rpm (67 Hz) and tight clearances, they are not very tolerant of solids such as sand.

Cables

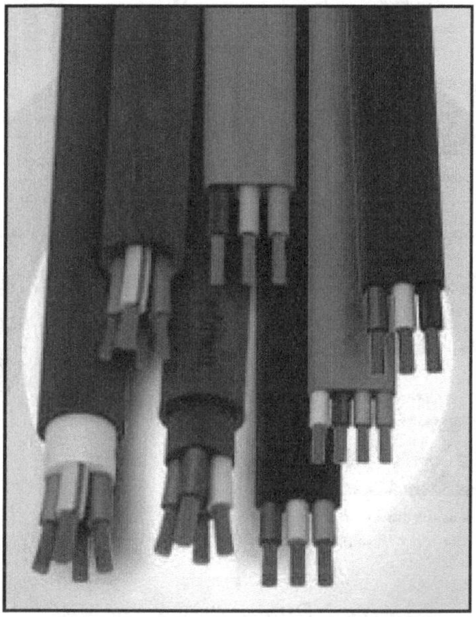

Figure: *Submersible Pump Cables: 3&4 Core Round and Flat Cables in PVC and Rubber Insulation*

Submersible pump cable are designed for use in wet ground or under water, with types specialized for pump environmental conditions.

A submersible pump cable is a specialized product to be used for a submersible pump in a deep well, or in similarly harsh conditions. The cable needed for this type of application must be durable and reliable as the installation location and environment can be extremely restrictive as well as hostile. As such, submersible pump cable can be used in both fresh and salt water. It is also suitable for direct burial and within well castings. A submersible pump cable's area of installation is physically restrictive. Cable manufacturers must keep these factors in mind to achieve the highest possible degree of reliability. The size and shape of submersible pump cable can vary depending on the usage and preference and pumping instrument of the installer. Pump cables are made in single and multiple conductor types and may be flat or round in cross section; some types include control wires as well as power conductors for the pump motor. Conductors are often colour-coded for identification and an overall cable jacket may also be colour-coded.

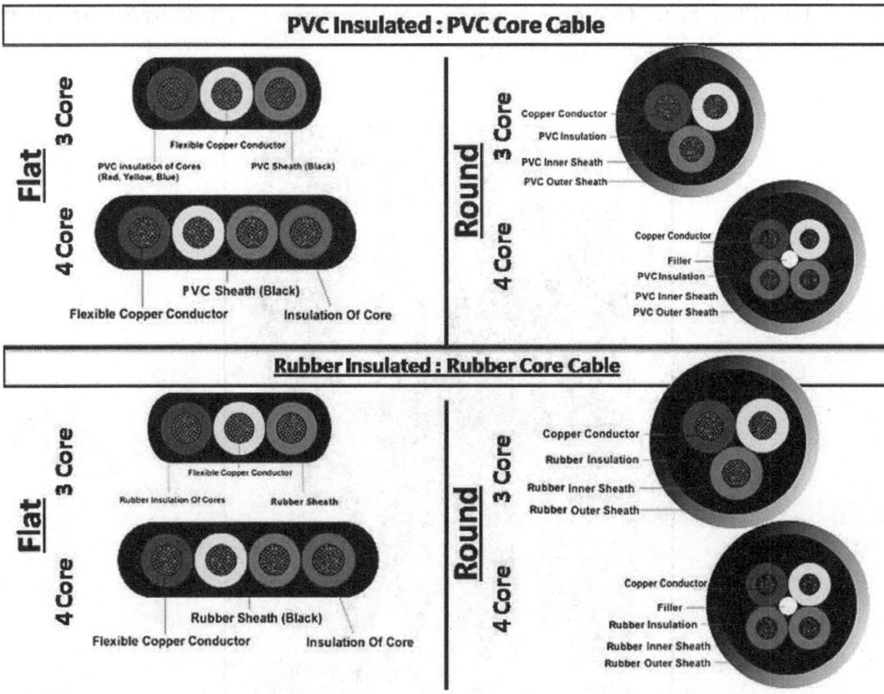

Figure: *Different types of SPC*

Major types of cable include:

In 3&4 Core cable as per right side *SPC types* image shown, Plain Copper/Tinned Copper used as conductor.

- PVC 3&4 Core Cable
  - o Flat Cable
  - o Round Cable
- Rubber 3&4 Core Cable
  - o Flat Cable
  - o Round Cable
- Flat Drincable
- HO7RN-F Cable.

Landing Nipple

A completion component fabricated as a short section of heavy wall tubular with a machined internal surface that provides a seal area and a locking profile. Landing nipples are included in most completions at predetermined intervals to enable the installation of flow-control devices, such as plugs and chokes. Three basic types of landing nipple are commonly used: no-go nipples, selective-landing nipples and ported or safety-valve nipples.

Sliding Sleeve

A sliding sleeve is a standard component for the completion of an oil or gas well. Their main uses are to shut off flow from one or more reservoir zones or to regulate pressure between zones.

There are two main categories of sliding sleeves: open/close and choking. Open/close sleeves are shifted between a full open position and a closed position. They are used to shut off flow from a zone for economic reasons or to shut off a zone that is depleting or producing too much water. In multi-zone wells, they are used to regulate which zones to produce from and which ones to shut off. Mechanically actuated sleeves are simple and inexpensive but require actuation by a "lock," which must be run in the well on wireline or coiled tubing. Hydraulically actuated sleeves are more complicated but can be actuated from a small pump at surface. King sleeves can be used to regulate the pressure between two or more zones. They are also used to regulate the flow of fluid into a well during proppant fracturing or hydraulic fracturing operations. Choking sleeves are all hydraulically actuated and have a much more complex design than open/close sleeves.

Operation

A separation tool is pinned in the open position when run by slickline. This provides a flow path through the centre of the toolwhile it is being set in the sliding sleeve. The separation tool adapts to most manufacturers' locks to match the nipple profile in the sliding sleeve. The lock and separation tool assembly are installed with the appropriate running tool and prong. When production is desired from an upper zone while blanking off the lower zone, a sliding sleeve with a nipple profile above and a polished sub below is installed in the tubing string opposite the upper zone. Packers are used to isolate the zones.

The sliding sleeve can be shifted to the open position and the separation tool can be run and locked into the bore of the sliding sleeve by standard slickline methods. The separation tool allows flow to enter from the annulus and produce up the tubing. The tool blanks off the tubing below the sleeve to isolate the lower zone.

The separation tool assembly is run on an appropriate running tool complete with a running prong. The running prong is connected to the top of the internal seal prong with a shear pin, and the running tool is attached to the lock in the normal manner. After locating the sliding sleeve, the operator jars downward to set the lock, then pulls upward to close the isolation plug in the separation tool. As this happens, a garter spring in the isolation plug moves a steel shear pin into a groove on the sealing prong, and an 0-ring on the internal prong seals to isolate the tubing below the sliding sleeve. The running prong that is pinned to the top of the internal prong is then sheared, and the slickline toolstring may be withdrawn from the well.

Production Packer

A production packer is a standard component of the completion hardware of oil or gas wells used to provide a seal between the outside of the production tubing and the inside of the casing, liner, or wellbore wall. Based on its primary use, packers can be divided into two main categories: production packers and service packers. Production packers are those that remain in the well during well production. Service packers are used temporarily during well service activities such as cement squeezing, acidizing, fracturing and well testing.

It is usually run in close to the bottom end of the production tubing and set at a point above the top perforations or sand screens. In wells with multiple reservoir zones, packers are used to isolate the perforations for each zone. In these situations, a sliding sleeve would be used to select which zone to produce. Packers may also be used to protect the casing from pressure and produced fluids, isolate sections of corroded casing, casing leaks or squeezed perforations, and isolate or temporarily abandon producing zones. In waterflooding developments in which water is injected into the reservoir, packers are used in injection wells to isolate the zones into which the water must be injected.

There are occasions in which running a packer may not be desirable. For example, high volume wells that are produced both up the tubing and annulus will not include a packer. Rod pumped wells are not normally run with packers because the associated gas is produced up the annulus. In general, well completions may not incorporate a packer when the annular space is used as a production conduit. A production packer is designed to grip and seal against the casing ID. Gripping is accomplished with metal wedges called "slips." These components have sharpened, carburized teeth that dig into the metal of the casing. Sealing is accomplished with large, cylindrical rubber elements. In situations where the sealed pressure is very high (above 5,000 psi), metal rings are used on either side of the elements to prevent the rubber from extruding.

A packer is run in the casing on production tubing or wireline. Once the desired depth is reached, the slips and element must be expanded out to contact the casing. Axial loads are applied to push the slips up a ramp and to compress the element, causing it to expand outward. The axial loads are applied either hydraulically, mechanically, or with a slow burning chemical charge. Most packers are "permanent" and require milling in order to remove them from the casing.

The main advantages of permanent packers are lower cost and greater sealing and gripping capabilities. In situations where a packer must be easily removed from the well, such as secondary recoveries, re-completions, or to change out the production tubing, a retrievable packer must be used. To unset the tool, either a metal ring is sheared or a sleeve is shifted to disengage connecting components. Retrievable packers have a more complicated design and generally lower sealing and gripping capabilities, but after removal and subsequent servicing, they can be reused.

Packer Applications: Casing Protection Separation of multiple Zones Isolation Packers Elimination of Surging and heading Sub-surface safety control Artificial/Gas Lift There are three types of packers; mechanical and hydraulic set and permanent. All packers fall into one or a combination of these.

Mechanical Set Packers: These are set by some form of tubing movement, usually a rotation or upward /downward motion.

Others can be weight set: The tubing weight can be used to compress and expand the sealing element. By a simple up string pull the packer is released. It is used best in shallow low pressure wells that are straight. It is not designed to withstand pressure differences unless a hydraulic hold down is incorporated.

Tension Set packers: They are set by pulling a tension on the tubing, slacking off releases the packer. Good for shallow wells with moderate pressure differences. The lower pressure helps to increase the setting force on the packer. Used in a stimulation well.

Rotation Set packer: Tubing rotation is used to set the packer to mechanically lock it in; an left hand turn engages and a right hand turn retrieves it.

Hydraulic Set packer: They use fluid pressure to drive the cone behind the slips. Once set they remain set by the use of either entrapped pressure or a mechanical lock. They are released by picking up the tubing. They are good for used in deviated/ crooked holes where tubing movement is restricted or unwanted. The tubing can be hung in neutral tension.

Inflatable rubber/Balloon packers: Uses fluid pressure to inflate a balloon and set the packer. Used in wells where the casing or open holes are collapsed. They can't withstand high pressure differentials and are only used in specialty applications.

Permanent packers: Run and set on an electric wireline, drill pipe or tubing. Opposed slips are positioned to lock it in compression. Once set this packer is resistant to motion for either direction. Wireline uses an electric current to detonate an explosive charge to set the packer. A release stud then frees the assembly form the packer. Tubing can be used by applying rotation or a pull or a combination of both. They are good in wells that have high pressure differentials or large tubing load variations and can be set precisely. They can be set the deepest.

Cement packers: In this case the tubing is cemented in place inside the casing or open hole. This type of packer is cheap.

Factors Affecting Packers

Temperature and pressure can affect how the tubing and the packer behave as this could cause changes in the packer and tubing expansion rates. If the packer allows free motion then the tubing can elongate or shorten. If not the tensile and compressive forces can develop within.

Downhole Gauges

This is an electronic or fibre-optic sensor to provide continuous monitoring of downhole pressure and temperature. Gauges either use a 1/4" control line clamped onto the outside of the tubing string to provide an electrical or fibre-optic communication to surface, or transmit measured data to surface by acoustic signal in the tubing wall.

Perforated Joint

This is a length of tubing with holes punched into it. If used, it will normally be positioned below the packer and will offer an alternative entry path for reservoir fluids into the tubing in case the shoe becomes blocked, for example, by a stuck perforation gun.

Formation Isolation Valve

This component, placed towards the foot of the completion string, is used to provide two way isolation from the formation for completion operations without the need for kill weight fluids. Their use is sporadic as they do not enjoy the best reputation for reliability when it comes to opening them at the end of the completion process.

Centralizer

In highly deviated wells, this components may be included towards the foot of the completion. It consists of a large collar, which keeps the completion string centralised within the hole.

Wireline Entry Guide

This component is often installed at the end of the tubing, or "the shoe". It is intended to make pulling out wireline tools easier by offering a guiding surface for the toolstring to re-enter the tubing without getting caught on the side of the shoe.

- Schlumberger WEG (pdf) Last accessed: 15-Jun-2011

Perforating and Stimulating

In cased hole completions (the majority of wells), once the completion string is in place, the final stage is to make a connection between the wellbore and the formation. This is done by running perforation guns to blast holes in the casing or liner to make a connection. Modern perforations are made using shaped explosive charges, similar to the armor-penetrating charge used on antitank rockets (bazookas). Sometimes once the well is fully completed, further stimulation is necessary to achieve the planned productivity. There are a number of stimulation techniques.

Acidizing

This involves the injection of chemicals to eat away at any skin damage, "cleaning up" the formation, thereby improving the flow of reservoir fluids. A strong acid (usually HCl) is used to dissolve rock formations, but this acid does not react with the Hydrocarbons. As a result the Hydrocarbons are more accessible. Acid can also be used to clean the wellbore of some scales that form from mineral laden produced water.

Fracturing

This means creating and extending fractures from the perforation tunnels deeper into the formation, increasing the surface area for formation fluids to flow into the well, as well as extending past any possible damage near the wellbore. This may be done by injecting fluids at high pressure (hydraulic fracturing), injecting fluids laced with round granular material (proppant fracturing), or using explosives to generate a high pressure and high speed gas flow (TNT or PETN up to 1,900,000 psi (13,000,000 kPa)) and (propellant stimulation up to 4,000 psi (28,000 kPa)).

Acidizing and Fracturing (Combined Method)

This involves use of explosives and injection of chemicals to increase acid-rock contact.

Nitrogen Circulation

Sometimes, productivity may be hampered due to the residue of completion fluids, heavy brines, in the wellbore. This is particularly a problem in gas wells. In these cases, coiled tubing may be used to pump nitrogen at high pressure into the bottom of the borehole to circulate out the brine.

Production

The production stage is the most important stage of a well's life, when the oil and gas are produced. By this time, the oil rigs and workover rigs used to drill and complete the well have moved off the wellbore, and the top is usually outfitted with a collection of valves called a Christmas tree or production tree. These valves regulate pressures, control flows, and allow access to the wellbore in case further completion work is needed. From the outlet valve of the production tree, the flow can be connected to a distribution network of pipelines and tanks to supply the product to refineries, natural gas compressor stations, or oil export terminals.

As long as the pressure in the reservoir remains high enough, the production tree is all that is required to produce the well. If the pressure depletes and it is considered economically viable, an artificial lift method mentioned in the completions section can be employed.

Workovers are often necessary in older wells, which may need smaller diameter tubing, scale or paraffin removal, acid matrix jobs, or completing new zones of interest in a shallower reservoir. Such remedial work can be performed using workover rigs – also known as *pulling units, completion rigs* or "service rigs" – to pull and replace tubing, or by the use of well intervention techniques utilizing coiled tubing. Depending on the type of lift system and wellhead a rod rig or flushby can be used to change a pump without pulling the tubing.

Enhanced recovery methods such as water flooding, steam flooding, or CO_2 flooding may be used to increase reservoir pressure and provide a "sweep" effect to push hydrocarbons out of the reservoir.

Such methods require the use of injection wells (often chosen from old production wells in a carefully determined pattern), and are used when facing problems with reservoir pressure depletion, high oil viscosity, or can even be employed early in a field's life.

In certain cases – depending on the reservoir's geomechanics – reservoir engineers may determine that ultimate recoverable oil may be increased by applying a waterflooding strategy early in the field's development rather than later. Such enhanced recovery techniques are often called "tertiary recovery".

Abandonment

A well is said to reach an "economic limit" when its most efficient production rate does not cover the operating expenses, including taxes.

The economic limit for oil and gas wells can be expressed using these formula:

Oil fields:

$$EL_{oil} = \frac{WI \times LOE}{NRI[P_o + (P_g \times GOR)/1{,}000] \times (1-T)}$$

Gas fields:

$$EL_{gas} = \frac{WI \times LOE}{NRI[(P_o \times Y) + P_g] \times (1-T)}$$

Where:

EL_{oil} is an oil well's economic limit in oil barrels per month (bbls/month).

EL_{gas} is a gas well's economic limit in thousand standard cubic feet per month (MSCF/month).

P_o, P_g are the current prices of oil and gas in dollars per barrels and dollars per MSCF respectively.

LOE is the lease operating expenses in dollars per well per month.

WI working interest, as a fraction.

NRI net revenue interest, as a fraction.

GOR gas/oil ratio as bbls/MSCF.

y condensate yield as barrel/million standard cubic feet.

T production and severance taxes, as a fraction.

When the economic limit is raised, the life of the well is shortened and proven oil reserves are lost. Conversely, when the economic limit is lowered, the life of the well is lengthened.

When the economic limit is reached, the well becomes a liability and is abandoned. In this process, tubing is removed from the well and sections of well bore are filled with concrete to isolate the flow path between gas and water zones from each other, as well as the surface.

Completely filling the well bore with concrete is costly and unnecessary. The surface around the wellhead is then excavated, and the wellhead and casing are cut off, a cap is welded in place and then buried.

At the economic limit there often is still a significant amount of unrecoverable oil left in the reservoir. It might be tempting to defer physical abandonment for an extended period of time, hoping that the oil price will go up or that new supplemental recovery techniques will be perfected.

In these cases, temporary plugs will be placed downhole and locks attached to the wellhead to prevent tampering. Thousands of "abandoned" wells throughout North America still exist in this state, waiting to see what the market will do before "permanent" abandonment. Often, lease provisions and governmental regulations usually require quick abandonment; liability and tax concerns also may favour abandonment.

In theory an abandoned well can be reentered and restored to production (or converted to injection service for supplemental recovery or for downhole hydrocarbons storage), but reentry often proves to be difficult mechanically and not cost effective.

Types of Wells

Fossil-fuel wells come in many varieties. By produced fluid, there can be wells that produce oil, wells that produce oil *and* natural gas, or wells that *only* produce natural gas. Natural gas is almost always a byproduct of producing oil, since the small, light gas carbon chains come out of solution as they undergo pressure reduction from the reservoir to the surface, similar to uncapping a bottle of soda pop where the carbon dioxide effervesces.

Unwanted natural gas can be a disposal problem at the well site. If there is not a market for natural gas near the wellhead it is virtually valueless since it must be piped to the end user.

Figure: *A natural gas well in the southeast Lost Hills Field, California, US.*

Until recently, such unwanted gas was burned off at the wellsite, but due to environmental concerns this practice is becoming less common.

Often, unwanted (or 'stranded' gas without a market) gas is pumped back into the reservoir with an 'injection' well for disposal or repressurizing the producing formation. Another solution is to export the natural gas as a liquid. Gas to liquid, (GTL) is a developing technology that converts stranded natural gas into synthetic gasoline, diesel or jet fuel through the Fischer-Tropsch process developed in World War II Germany.

Such fuels can be transported through conventional pipelines and tankers to users. Proponents claim GTL fuels burn cleaner than comparable petroleum fuels.

Most major international oil companies are in advanced development stages of GTL production, e.g. the 140,000 bbl/d (22,000 m^3/d) Pearl GTL plant in Qatar, scheduled to come online in 2011. In locations such as the United States with a high natural gas demand, pipelines are constructed to take the gas from the wellsite to the end consumer.

Figure: *Raising the stern*

Another obvious way to classify oil wells is by land or offshore wells. There is very little difference in the well itself.

An offshore well targets a reservoir that happens to be underneath an ocean. Due to logistics, drilling an offshore well is far more costly than an onshore well. By far the most common type is the onshore well.

These wells dot the Southern and Central Great Plains, Southwestern United States, and are the most common wells in the Middle East.

Another way to classify oil wells is by their purpose in contributing to the development of a resource. They can be characterized as:

- *wildcat wells* are those drilled outside of and not in the vicinity of known oil or gas fields.
- *exploration wells* are drilled purely for exploratory (information gathering) purposes in a new area.
- *appraisal wells* are used to assess characteristics (such as flow rate) of a proven hydrocarbon accumulation.
- *production wells* are drilled primarily for producing oil or gas, once the producing structure and characteristics are determined.

Figure: *Oil extraction in Boryslav in 1909*

At a producing well site, active wells may be further categorised as:

- *oil producers* producing predominantly liquid hydrocarbons, but mostly with some associated gas.
- *gas producers* producing almost entirely gaseous hydrocarbons.
- *water injectors* injecting water into the formation to maintain reservoir pressure, or simply to dispose of water produced with the hydrocarbons because even after treatment, it would be too oily and too saline to be considered clean for dumping overboard offshore, let alone into a fresh water resource in the case of onshore wells. Water injection into the producing zone frequently has an element of reservoir management; however, often produced water disposal is into shallower zones safely beneath any fresh water zones.
- *aquifer producers* intentionally producing water for re-injection to manage pressure. If possible this water will come from the reservoir itself. Using aquifer produced water rather than water from other sources is to preclude chemical incompatibility that might lead to reservoir-plugging precipitates. These wells will generally be needed only if produced water from the oil or gas producers is insufficient for reservoir management purposes.
- *gas injectors* injecting gas into the reservoir often as a means of disposal or sequestering for later production, but also to maintain reservoir pressure.

Lahee Classification

- *New Field Wildcat* (NFW) – far from other producing fields and on a structure that has not previously produced.
- *New Pool Wildcat* (NPW) – new pools on already producing structure.
- *Deeper Pool Test* (DPT) – on already producing structure and pool, but on a deeper pay zone.
- *Shallower Pool Test* (SPT) – on already producing structure and pool, but on a shallower pay zone.
- *Outpost* (OUT) – usually two or more locations from nearest productive area.
- *Development Well* (DEV) – can be on the extension of a pay zone, or between existing wells (*Infill*).

Cost

The cost of a well depends mainly on the daily rate of the drilling rig, the extra services required to drill the well, the duration of the well programme (including downtime and weather time), and the remoteness of the location (logistic supply costs). The daily rates of offshore drilling rigs vary by their capability, and the market availability. Rig rates reported by industry web service show that the deepwater water floating drilling rigs are over twice that of the shallow water fleet, and rates for jackup fleet can vary by factor of 3 depending upon capability. With deepwater drilling rig rates in 2010 of around $420,000/day, and similar additional spread costs, a deep water well of duration of 100 days can cost around US$100 million. With high performance jackup rig rates in 2010 of around $150,000, and similar service costs, a high pressure, high temperature well of duration 100 days can cost about US$30 million. Onshore wells can be considerably cheaper, particularly if the field is at a shallow depth, where costs range from less than $1 million to $15 million for deep and difficult wells.

The total cost of an oil well mentioned does not include the costs associated with the risk of explosion and leakage of oil. Those costs include the cost of protecting against such disasters, the cost of the cleanup effort, and the hard-to-calculate cost of damage to the company's image. Offshore platforms (the structure supporting the wells) often provide habitat for marine life. After the wells have been abandoned, sometimes the platforms can be toppled in place or moved elsewhere to be dropped to the ocean floor to produce artificial reefs.

Chapter 4
Properties of Testing

Crude Oil Properties

Some of the most important physical properties of crude oil include:

°API

°API = (141.5/SG $_{15^oF}$) – 131.5 The purpose of this equation was to extend the range of the specific gravity scale. Crude oil SG changes, although small, may be important.

Crude Oils °API = 10 – 50

Higher °API, more paraffinic crude, higher yields of gasoline.

Lower °API, more aromatic crude, lower yields of gasoline.

Viscosity

Resistance to flow, usually measured @ 100°F in centistokes (kinimatic viscosity)

Pour Point

Measured by ASTM D-97 – temperature at which oil ceases to flow. Diesel may contain waxes, smaller than candle wax, which could solidify in cold weather.

Flash Point

Temperature above which the oil will spontaneously combust. Fractions in vacuum tower are the least combustible. They are the heaviest.

Vapor Pressure

Measured by ASTM D-323. Also know as Reid vapor pressure (RVP). True vapor pressure is usually 5-9% > RVP

Carbon Residue

The solid residue (%wt) remaining after heating to coking temperatures (700-800°C)

ASTM D-524 Ramsbottom Carbon

ASTM D-189 Conradson Carbon

CCR incr. then Asphaltene incr.

Salt Content

Measured by ASTM – 3230 (lb NaCl/1000 bbl)

Desalting is necessary because NaCl content > 10 lbs/1000 bbl leads to corrosion

Metals

Measured by EPA Method 3040 These include Ni, V, Ag, Hg, Na, and Ca. Metals can cause catalyst deactivation and corrosion.

Sediment and Water

Measured by ASTM D – 96 These inorganic particles can lead to operational problems.

Acidity

Measured by ASTM – 664

Sulphur

Measured by ASTM D – 129, 1552, 2622

Sour crudes > 0.5 wt% and sweet crudes < 0.5 wt%. Today it is difficult to find crudes below 1% sulphur.

TBP Distillation Data

Butanes and lighter	55-175 °F
Light Gasoline	175-300 °F
Light naphtha	300-400 °F
Heavy naptha	400-500 °F
Kerosene	500-650 °F
Atmosphere Gas Oil	650-800 °F
Light Vacuum Gas Oil	800-1000 °F
Hvy. Vacuum Gas Oil	1000 °F
Vacuum Residue	> 1000 °F

Specifications and Environmental Regulations for Gasoline and Diesel

Engine knock is caused by the spontaneous ignition of fuel (higher octane gas reduces engine knock). Gasoline engines use spark ignition. Octane number is determined by (Research Octane + Motor Octane)/2.

Research octane is measured at low rpm, characteristic of city driving. Motor octane is measured at high rpm, characteristic of highway driving. The octane #s are reported by comparison of the gasoline's performance to that of a binary mixture of:

n-heptane (0)

2,2,4 – trimethylpentane (100)

Octane Rating	***Fuel Type***	***Compression Ratio***
87	regular	9
89	plus	
93	super	12

Octane # of straight run crude oil is ~ 40

EPA regulations limiting benzene to 1%, aromatics to 10%, and sulphur to 30 ppm

Diesel fuel is rated by a cetane number. The desirable range for the cetane number is between 40-50

EPA regulations limiting sulphur content to 50 ppm in diesel.

Petroleum Properties

Composition: In its strictest sense, petroleum includes only crude oil, but in common usage it includes all liquid, gaseous, and solid hydrocarbons. Under surface pressure and temperature conditions, lighter hydrocarbons methane, ethane, propane and butane occur as gases, while pentane and heavier ones are in the form of liquids or solids. However, in an underground oil reservoir the proportions of gas, liquid, and solid depend on subsurface conditions and on the phase diagram of the petroleum mixture.

An oil well produces predominantly crude oil, with some natural gas dissolved in it. Because the pressure is lower at the surface than underground, some of the gas will come out of solution and be recovered (or burned) as *associated gas* or *solution gas*. A gas well produces predominantly natural gas. However, because the underground temperature and pressure are higher than at the surface, the gas may contain heavier hydrocarbons such as pentane, hexane, and heptane in the gaseous state. At surface conditions these will condense out of the gas to form natural gas condensate, often shortened to *condensate*. Condensate resembles petrol in appearance and is similar in composition to some volatile light crude oils.

The proportion of light hydrocarbons in the petroleum mixture varies greatly among different oil fields, ranging from as much as 97 percent by weight in the lighter oils to as little as 50 percent in the heavier oils and bitumens.

The hydrocarbons in crude oil are mostly alkanes, cycloalkanes and various aromatic hydrocarbons while the other organic compounds contain nitrogen, oxygen and sulphur, and trace amounts of metals such as iron, nickel, copper and vanadium. The exact molecular composition varies widely from formation to formation but the proportion of chemical elements vary over fairly narrow limits as follows:

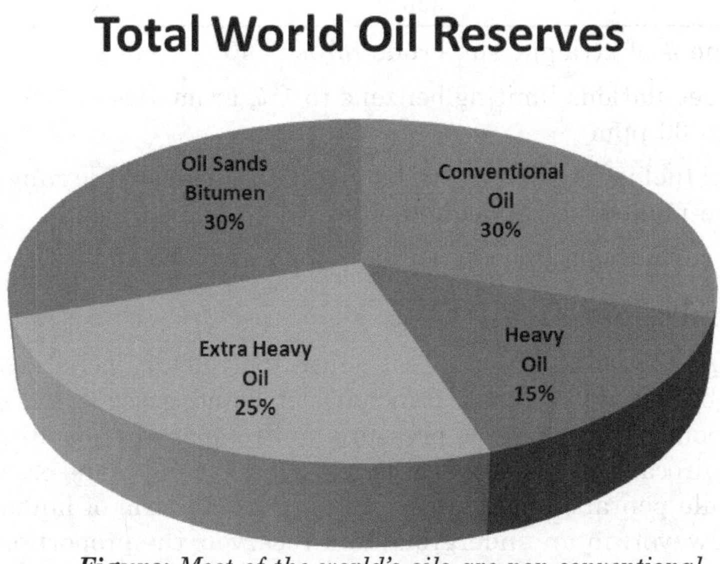

Figure: *Most of the world's oils are non-conventional.*

Composition by Weight	
Element	Percent range
Carbon	83 to 87%
Hydrogen	10 to 14%
Nitrogen	0.1 to 2%
Oxygen	0.05 to 1.5%
Sulphur	0.05 to 6.0%
Metals	< 0.1%

Four different types of hydrocarbon molecules appear in crude oil. The relative percentage of each varies from oil to oil, determining the properties of each oil.

Composition by weight		
Hydrocarbon	***Average***	***Range***
Alkanes (paraffins)	30%	15 to 60%
Naphthenes	49%	30 to 60%
Aromatics	15%	3 to 30%
Asphaltics	6%	remainder

Crude oil varies greatly in appearance depending on its composition. It is usually black or dark brown (although it may be yellowish, reddish, or even greenish). In the reservoir it is usually found in association with natural gas, which being lighter forms a gas cap over the petroleum, and saline water which, being heavier than most forms of crude oil, generally sinks beneath it. Crude oil may also be found in semi-solid form mixed with sand and water, as in the Athabasca oil sands in Canada, where it is usually referred to as crude bitumen.

In Canada, bitumen is considered a sticky, black, tar-like form of crude oil which is so thick and heavy that it must be heated or diluted before it will flow. Venezuela also has large amounts of oil in the Orinoco oil sands, although the hydrocarbons trapped in them are more fluid than in Canada and are usually called extra heavy oil. These oil sands resources are called unconventional oil to distinguish them from oil which can be extracted using traditional oil well methods. Between them, Canada and Venezuela contain an estimated 3.6 trillion barrels (570×10^9 m^3) of bitumen and extra-heavy oil, about twice the volume of the world's reserves of conventional oil.

Petroleum is used mostly, by volume, for producing fuel oil and petrol, both important *"primary energy"* sources. 84 percent by volume of the hydrocarbons present in petroleum is converted into energy-rich fuels (petroleum-based fuels), including petrol, diesel, jet, heating, and other fuel oils, and liquefied petroleum gas. The lighter grades of crude oil produce the best yields of these products, but as the world's reserves of light and medium oil are depleted, oil refineries are increasingly having to process heavy oil and bitumen, and use more complex and expensive methods to produce the products required. Because heavier crude oils have too much carbon and not enough hydrogen, these processes generally involve removing carbon from or adding hydrogen to the molecules, and using fluid catalytic cracking to convert the longer, more complex molecules in the oil to the shorter, simpler ones in the fuels.

Due to its high energy density, easy transportability and relative abundance, oil has become the world's most important source of energy since the mid-1950s. Petroleum is also the raw material for many chemical products, including pharmaceuticals, solvents, fertilizers, pesticides, and plastics; the 16 percent not used for energy production is converted into these other materials. Petroleum is found in porous rock formations in the upper strata of some areas of the Earth's crust. There is also petroleum in oil sands (tar sands). Known oil reserves are typically estimated at around 190 km^3 (1.2 trillion (short scale) barrels) without oil sands, or 595 km^3 (3.74 trillion barrels) with oil sands. Consumption is currently around 84 million barrels (13.4×10^6 m^3) per day, or 4.9 km^3 per year. Which in turn yields a remaining oil supply of only about 120 years, if current demand remain static.

Chemistry

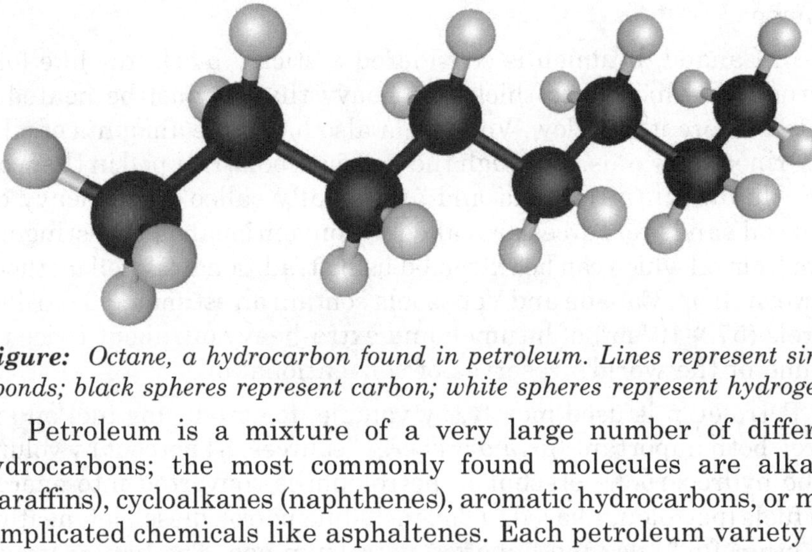

Figure: *Octane, a hydrocarbon found in petroleum. Lines represent single bonds; black spheres represent carbon; white spheres represent hydrogen.*

Petroleum is a mixture of a very large number of different hydrocarbons; the most commonly found molecules are alkanes (paraffins), cycloalkanes (naphthenes), aromatic hydrocarbons, or more complicated chemicals like asphaltenes. Each petroleum variety has a unique mix of molecules, which define its physical and chemical properties, like colour and viscosity.

The *alkanes*, also known as *paraffins*, are saturated hydrocarbons with straight or branched chains which contain only carbon and hydrogen and have the general formula C_nH_{2n+2}. They generally have from 5 to 40 carbon atoms per molecule, although trace amounts of shorter or longer molecules may be present in the mixture.

The alkanes from pentane (C_5H_{12}) to octane (C_8H_{18}) are refined into petrol, the ones from nonane (C_9H_{20}) to hexadecane ($C_{16}H_{34}$) into diesel fuel, kerosene and jet fuel. Alkanes with more than 16 carbon atoms can be refined into fuel oil and lubricating oil. At the heavier end of the range, paraffin wax is an alkane with approximately 25 carbon atoms, while asphalt has 35 and up, although these are usually cracked by modern refineries into more valuable products. The shortest molecules, those with four or fewer carbon atoms, are in a gaseous state at room temperature. They are the petroleum gases. Depending on demand and the cost of recovery, these gases are either flared off, sold as liquified petroleum gas under pressure, or used to power the refinery's own burners. During the winter, butane (C_4H_{10}), is blended into the petrol pool at high rates, because its high vapor pressure assists with cold starts. Liquified under pressure slightly above atmospheric, it is best known for powering cigarette lighters, but it is also a main fuel source for many developing countries. Propane can be liquified under modest pressure, and is consumed for just about every application relying on petroleum for energy, from cooking to heating to transportation.

The *cycloalkanes*, also known as *naphthenes*, are saturated hydrocarbons which have one or more carbon rings to which hydrogen atoms are attached according to the formula C_nH_{2n}. Cycloalkanes have similar properties to alkanes but have higher boiling points.

The *aromatic hydrocarbons* are unsaturated hydrocarbons which have one or more planar six-carbon rings called benzene rings, to which hydrogen atoms are attached with the formula C_nH_n. They tend to burn with a sooty flame, and many have a sweet aroma. Some are carcinogenic.

These different molecules are separated by fractional distillation at an oil refinery to produce petrol, jet fuel, kerosene, and other hydrocarbons. For example, 2,2,4-trimethylpentane (isooctane), widely used in petrol, has a chemical formula of C_8H_{18} and it reacts with oxygen exothermically:

$$2C8H18_{(l)} + 25O2_{(g)} \rightarrow 16CO2_{(g)} + 18H2O_{(g)} \left(\Delta H = -10.86 \text{ MJ / mol of octane}\right)$$

The amount of various molecules in an oil sample can be determined in laboratory. The molecules are typically extracted in a solvent, then separated in a gas chromatograph, and finally determined with a suitable detector, such as a flame ionization detector or a mass spectrometer. Due to the large number of co-eluted hydrocarbons within oil, many cannot be resolved by traditional gas chromatography and typically appear as a hump in the chromatogram. This unresolved complex mixture (UCM) of hydrocarbons is particularly apparent when analysing weathered oils and extracts from tissues of organisms exposed to oil.

Incomplete combustion of petroleum or petrol results in production of toxic byproducts. Too little oxygen results in carbon monoxide. Due to the high temperatures and high pressures involved, exhaust gases from petrol combustion in car engines usually include nitrogen oxides which are responsible for creation of photochemical smog.

Empirical Equations for Thermal Properties

Heat of Combustion: At a constant volume the heat of combustion of a petroleum product can be approximated as follows:

$$Q_v = 12400, -2{,}100d^2 .$$

where Q_v is measured in cal/gram and d is the specific gravity at 60 °F (16 °C).

Thermal conductivity: The thermal conductivity of petroleum based liquids can be modelled as follows:

$$K = \frac{0.823}{d}[1 - 0.0203(t - 32)]\,0.547$$

where K is measured in BTU · $hr^{-1}ft^{-2}$, t is measured in F and d is the specific gravity at 60 °F (16 °C).

Specific Heat

The specific heat of a petroleum oils can be modelled as follows:

$$c = \frac{1}{d}[0.388 + 0.00045t],$$

where c is measured in BTU/lbm-°F, t is the temperature in Fahrenheit and d is the specific gravity at 60 °F (16 °C).

In units of kcal/(kg·°C), the formula is:

$$c = \frac{1}{d}[0.4024 + 0.00081t],$$

where the temperature t is in Celsius and d is the specific gravity at 15 °C.

Latent Heat of Vaporization

The latent heat of vaporization can be modelled under atmospheric conditions as follows:

$$L = \frac{1}{d}[110.9 - 0.09t],$$

where L is measured in BTU/lbm, t is measured in °F and d is the specific gravity at 60 °F (16 °C).

In units of kcal/kg, the formula is:

$$L = \frac{1}{d}[194.4 - 0.162t],$$

where the temperature t is in Celsius and d is the specific gravity at 15 °C.

Formation

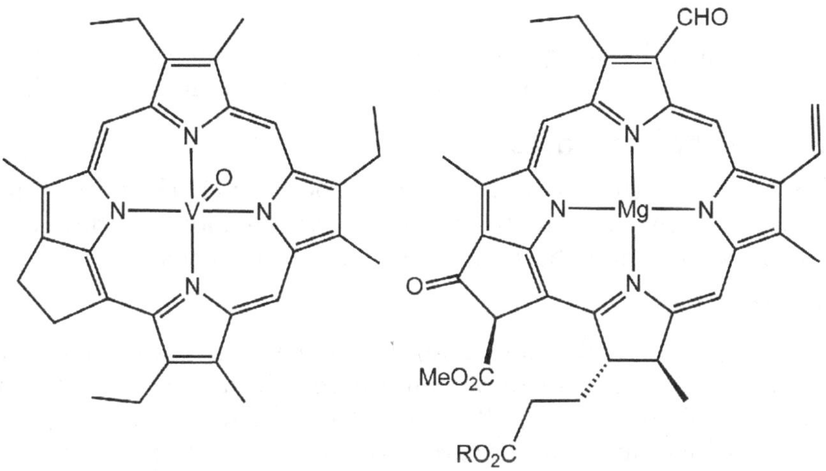

Figure: *Structure of a vanadium porphyrin compound (left) extracted from petroleum by Alfred E. Treibs, father of organic geochemistry. Treibs noted the close structural similarity of this molecule and chlorophyll a (right).*

Petroleum is a fossil fuel derived from ancient fossilized organic materials, such as zooplankton and algae. Vast quantities of these remains settled to sea or lake bottoms, mixing with sediments and being buried under anoxic conditions. As further layers settled to the sea or lake bed, intense heat and pressure built up in the lower regions. This process caused the organic matter to change, first into a waxy material known as kerogen, which is found in various oil shales around the world, and then with more heat into liquid and gaseous hydrocarbons via a process known as catagenesis. Formation of petroleum occurs from hydrocarbon pyrolysis in a variety of mainly endothermic reactions at high temperature and/or pressure.

There were certain warm nutrient-rich environments such as the Gulf of Mexico and the ancient Tethys Sea where the large amounts of organic material falling to the ocean floor exceeded the rate at which it could decompose. This resulted in large masses of organic material being buried under subsequent deposits such as shale formed from mud. This massive organic deposit later became heated and transformed under pressure into oil.

Geologists often refer to the temperature range in which oil forms as an "oil window"—below the minimum temperature oil remains trapped in the form of kerogen, and above the maximum temperature the oil is converted to natural gas through the process of thermal cracking. Sometimes, oil formed at extreme depths may migrate and become trapped at a much shallower level. The Athabasca Oil Sands are one example of this.

An alternative mechanism proposed by Russian scientists in the mid 1850s is called the Abiogenic petroleum origin is a hypothesis.

The Properties of Gases

Gases have three characteristic properties: (1) they are easy to compress, (2) they expand to fill their containers, and (3) they occupy far more space than the liquids or solids from which they form.

Compressibility

An internal combustion engine provides a good example of the ease with which gases can be compressed. In a typical four-stroke engine, the piston is first pulled out of the cylinder to create a partial vacuum, which draws a mixture of gasoline vapor and air into the cylinder. The piston is then pushed into the cylinder, compressing the gasoline/air mixture to a fraction of its original volume.

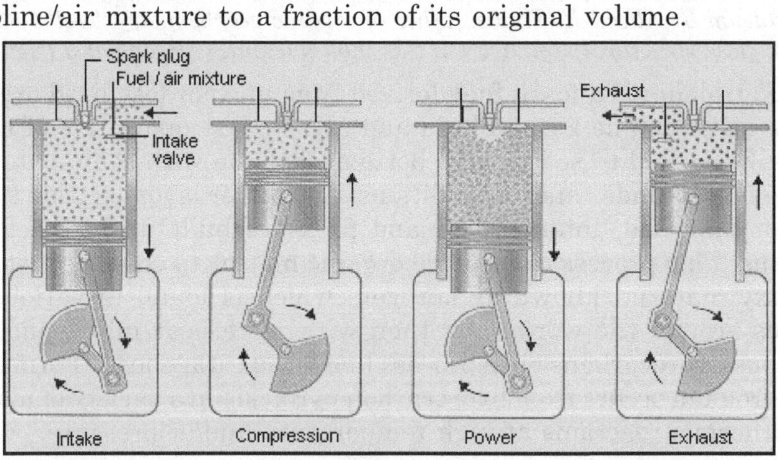

Figure: *The operation of a four-stroke engine can be divided into four cycles: intake, compression, power, and exhaust stages.*

The ratio of the volume of the gas in the cylinder after the first stroke to its volume after the second stroke is the *compression ratio* of the engine. Modern cars run at compression ratios of about 9:1, which means the gasoline-air mixture in the cylinder is compressed by a factor of nine in the second stroke. After the gasoline/air mixture is compressed, the spark plug at the top of the cylinder fires and the resulting explosion pushes the piston out of the cylinder in the third stroke. Finally, the piston is pushed back into the cylinder in the fourth stroke, clearing out the exhaust gases.

Liquids are much harder to compress than gases. They are so hard to compress that the hydraulic brake systems used in most cars operate on the principle that there is essentially no change in the volume of the brake fluid when pressure is applied to this liquid. Most solids are even harder to compress. The only exceptions belong to a rare class of compounds that includes natural and synthetic rubber. Most rubber balls that seem easy to compress, such as a racquetball, are filled with air, which is compressed when the ball is squeezed.

Expandability

Anyone who has walked into a kitchen where bread was baking has experienced the fact that gases expand to fill their containers, as the air in the kitchen becomes filled with wonderful odors. Unfortunately the same thing happens when someone breaks open a rotten egg and the characteristic odor of hydrogen sulfide (H_2S) rapidly diffuses through the room. Because gases expand to fill their containers, it is safe to assume that the volume of a gas is equal to the volume of its container.

Volumes of Gases Versus Volumes of Liquids or Solids

The difference between the volume of a gas and the volume of the liquid or solid from which it forms can be illustrated with the following examples. One gram of liquid oxygen at its boiling point (-183°C) has a volume of 0.894 mL. The same amount of O_2 gas at 0°C and atmospheric pressure has a volume of 700 mL, which is almost 800 times larger. Similar results are obtained when the volumes of solids and gases are compared. One gram of solid CO_2 has a volume of 0.641 mL. At 0°C and atmospheric pressure, the same amount of CO_2 gas has a volume of 556 mL, which is more than 850 times as large. As a general rule, the volume of a liquid or solid increases by a factor of about 800 when it forms a gas.

The consequences of this enormous change in volume is frequently used to do work. The steam engine, which brought about the industrial revolution, is based on the fact that water boils to form a gas (steam) that has a much larger volume. The gas therefore escapes from the container in which it was generated and the escaping steam can be made to do work. The same principle is at work when dynamite is used to blast rocks. In 1867, the Swedish chemist Alfred Nobel discovered that the highly dangerous liquid explosive known as nitroglycerin could be absorbed onto clay or sawdust to produce a solid that was much more stable and therefore safer to use. When dynamite is detonated, the nitroglycerin decomposes to produce a mixture of CO_2, H_2O, N_2, and O_2 gases.

$$4\,C_3H_5N_3O_9\,(l) \rightarrow 12\,CO_2\,(g) + 10\,H_2O(g) + 6\,N_2\,(g) + O_2\,(g)$$

Because 29 moles of gas are produced for every four moles of liquid that decompose, and each mole of gas occupies a volume roughly 800 times larger than a mole of liquid, this reaction produces a shock wave that destroys anything in its vicinity.

The same phenomenon occurs on a much smaller scale when we pop popcorn. When kernels of popcorn are heated in oil, the liquids inside the kernel turn into gases. The pressure that builds up inside the kernel is enormous and the kernel eventually explodes.

Pressure Versus Force

The volume of a gas is one of its characteristic properties. Another characteristic property is the pressure the gas exerts on its surroundings. Many of us got our first exposure to the pressure of a gas when we rode to the neighbourhood gas station to check the pressure of our bicycle tires. Depending on the kind of bicycle we had, we added air to the tires until the pressure gauge read between 30 and 70 pounds per square inch (lb/in^2 or psi). Two important properties of pressure can be obtained from this example.

1. The pressure of a gas becomes larger as more gas is added to the container.
2. Pressure is measured in units (such as lb/in^2) that describe the force exerted by the gas divided by the area over which this force is distributed.

The first conclusion can be summarized in the following relationship, where P is the pressure of the gas and n is the amount of gas in the container.

$$P \propto n$$

Because the pressure increases as gas is added to the container, P is directly proportional to n.

The second conclusion describes the relationship between pressure and force. Pressure is defined as the force exerted on an object divided by the area over which the force is distributed.

$$\Pr essure = \frac{force}{area}$$

The difference between pressure and force can be illustrated with an analogy based on a 10-penny nail, a hammer, and a piece of wood. By resting the nail on its point, and hitting the head with the hammer, we can drive the nail into the wood. But what happens if we turn the nail over and rest the head of the nail against the wood? If we hit the nail with the same force, we can't get the nail to stick into the wood.

When we hit the nail on the head, the force of this blow is applied to the very small area of the wood in contact with the sharp point of the nail, and the nail slips easily into the wood. But when we turn the nail over, and hit it on the point, the force is distributed over a much larger area. The force is now distributed over the surface of the wood that touches any part of the nail head. As a result, the pressure applied to the wood is much smaller and the nail just bounces off the wood.

Properties of Water

Water H_2O) is the most abundant compound on Earth's surface, covering about 70 percent of the planet. In nature, water exists in liquid, solid, and gaseous states. It is in dynamic equilibrium between the liquid and gas states at standard temperature and pressure. At room temperature, it is a tasteless and odorless liquid, nearly colourless with a hint of blue. Many substances dissolve in water and it is commonly referred to as *the universal solvent*. Because of this, water in nature and in use is rarely pure and some of its properties may vary slightly from those of the pure substance. However, there are also many compounds that are essentially, if not completely, insoluble in water. Water is the only common substance found naturally in all three common states of matter and it is essential for all life on Earth. Water usually makes up 55% to 78% of the human body.

In keeping with the basic rules of chemical nomenclature, water would have a systematic name of *dihydrogen monoxide*, but this is not among the names published by the International Union of Pure and Applied Chemistry and, rather than being used in a chemical context, the name is almost exclusively used as a humorous way to refer to water.

Forms of Water

Like many substances, water can take numerous forms that are broadly categorized by phase of matter. The liquid phase is the most common among water's phases (within the Earth's atmosphere and surface) and is the form that is generally denoted by the word "water." The solid phase of water is known as ice and commonly takes the structure of hard, amalgamated crystals, such as ice cubes, or loosely accumulated granular crystals, like snow. For a list of the many different crystalline and amorphous forms of solid H_2O. The gaseous phase of water is known as water vapor (or steam), and is characterized by water assuming the configuration of a transparent cloud. (Note that the visible steam and clouds are, in fact, water in the liquid form as minute droplets suspended in the air.) The fourth state of water, that of a supercritical fluid, is much less common than the other three and only rarely occurs in nature, in extremely uninhabitable conditions. When water achieves a specific critical temperature and a specific critical pressure (647 K and 22.064 MPa), liquid and gas phase merge to one homogeneous fluid phase, with properties of both gas and liquid. One example of naturally occurring supercritical water is found in the hottest parts of deep water hydrothermal vents, in which water is heated to the critical temperature by scalding volcanic plumes and achieves the critical pressure because of the crushing weight of the ocean at the extreme depths at which the vents are located. Additionally, anywhere there is volcanic activity below a depth of 2.25 km (1.40 mi) can be expected to have water in the supercritical phase.

Vienna Standard Mean Ocean Water is the current international standard for water isotopes. Naturally occurring water is almost completely composed of the neutron-less hydrogen isotope protium. Only 155 ppm include deuterium (^2H or D), a hydrogen isotope with one neutron, and fewer than 20 parts per quintillion include tritium (^3H or T), which has two.

Heavy water is water with a higher-than-average deuterium content, up to 100%. Chemically, it is similar but not identical to normal water. This is because the nucleus of deuterium is twice as heavy as protium, and this causes noticeable differences in bonding energies. Because water molecules exchange hydrogen atoms with one another, hydrogen deuterium oxide (DOH) is much more common in low-purity heavy water than pure dideuterium monoxide (D_2O). Humans are generally unaware of taste differences, but sometimes report a burning sensation or sweet flavour. Rats, however, are able to avoid heavy water by smell. Toxic to many animals, heavy water is used in the nuclear reactor industry to moderate (slow down) neutrons. Light water reactors are also common, where "light" simply designates normal water.

Light water more specifically refers to deuterium-depleted water (DDW), water in which the deuterium content has been reduced below the standard 155 ppm level.

Physics and Chemistry

Water is the chemical substance with chemical formula H_2O: one molecule of water has two hydrogen atoms covalently bonded to a single oxygen atom. Water is a tasteless, odorless liquid at ambient temperature and pressure, and appears colourless in small quantities, although it has its own intrinsic very light blue hue. Ice also appears colourless, and water vapor is essentially invisible as a gas.

Water is primarily a liquid under standard conditions, which is not predicted from its relationship to other analogous hydrides of the oxygen family in the periodic table, which are gases such as hydrogen sulfide. The elements surrounding oxygen in the periodic table, nitrogen, fluorine, phosphorus, sulphur and chlorine, all combine with hydrogen to produce gases under standard conditions. The reason that water forms a liquid is that oxygen is more electronegative than all of these elements with the exception of fluorine. Oxygen attracts electrons much more strongly than hydrogen, resulting in a net positive charge on the hydrogen atoms, and a net negative charge on the oxygen atom. The presence of a charge on each of these atoms gives each water molecule a net dipole moment. Electrical attraction between water molecules due to this dipole pulls individual molecules closer together, making it more difficult to separate the molecules and therefore raising the boiling point. This attraction is known as hydrogen bonding. The molecules of water are constantly moving in relation to each other, and the hydrogen bonds are continually breaking and reforming at timescales faster than 200 femtoseconds. However, this bond is sufficiently strong to create many of the peculiar properties of water, such as those that make it integral to life. Water can be described as a polar liquid that slightly dissociates disproportionately into the hydronium ion ($H_3O^+(aq)$) and an associated hydroxide ion ($OH^-(aq)$).

$$2H_2O(l) \rightleftharpoons H_3O+(aq) + OH^-(aq)$$

The dissociation constant for this dissociation is commonly symbolized as K_w and has a value of about 10^{-14} at 25 °C.

Percentage of elements in water by mass: 11.1% hydrogen, 88.9% oxygen.

The Self-diffusion coefficient of Water is $2.299 \cdot 10^{-9}\ m^2 \cdot s^{-1}$

Water, Ice and Vapor

Heat capacity and heats of vaporization and fusion:

Heat of vaporization	
Temperature (°C)	***H_v (kJ/mol)***
0	45.054
25	43.99
40	43.35
60	42.482
80	41.585
100	40.657
120	39.684
140	38.643
160	37.518
180	36.304
200	34.962
220	33.468
240	31.809
260	29.93
280	27.795
300	25.3
320	22.297
340	18.502
360	12.966
374	2.066

Water has a very high specific heat capacity – the second highest among all the heteroatomic species (after ammonia), as well as a high heat of vaporization (40.65 kJ/mol or 2257 kJ/kg at the normal boiling point), both of which are a result of the extensive hydrogen bonding between its molecules. These two unusual properties allow water to moderate Earth's climate by buffering large fluctuations in temperature. According to Josh Willis, of NASA's Jet Propulsion Laboratory, the oceans absorb one thousand times more heat than the atmosphere (air) and are holding 80 to 90% of the heat of global warming.

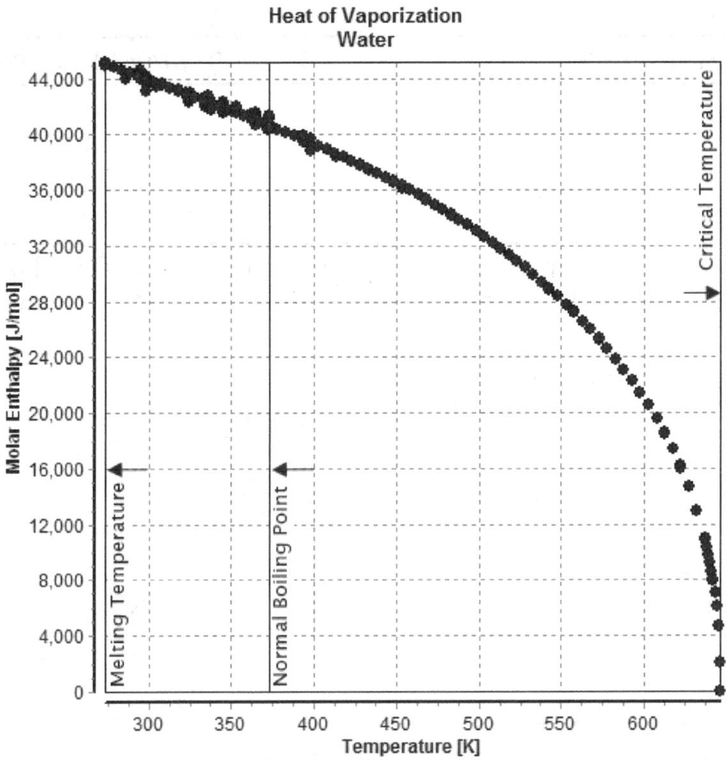

Figure: *Heat of vaporization of water from melting to critical temperature*

The specific enthalpy of fusion of water is 333.55 kJ/kg at 0 °C. Of common substances, only that of ammonia is higher. This property confers resistance to melting on the ice of glaciers and drift ice. Before and since the advent of mechanical refrigeration, ice was and still is in common use for retarding food spoilage.

Constant-pressure Heat Capacity

Temperature (°C)	C_p *(J/(g·K) at 100 kPa)*
0	4.2176
10	4.1921
20	4.1818
25	4.1814
30	4.1784
40	4.1785
50	4.1806
60	4.1843

Contd...

Temperature (°C)	C_p *(J/(g·K) at 100 kPa)*
70	4.1895
80	4.1963
90	4.205
100	4.2159

Note that the specific heat capacity of ice at –10 °C is about 2.05 J/(g·K) and that the heat capacity of steam at 100 °C is about 2.080 J/(g·K).

Density of Water and Ice

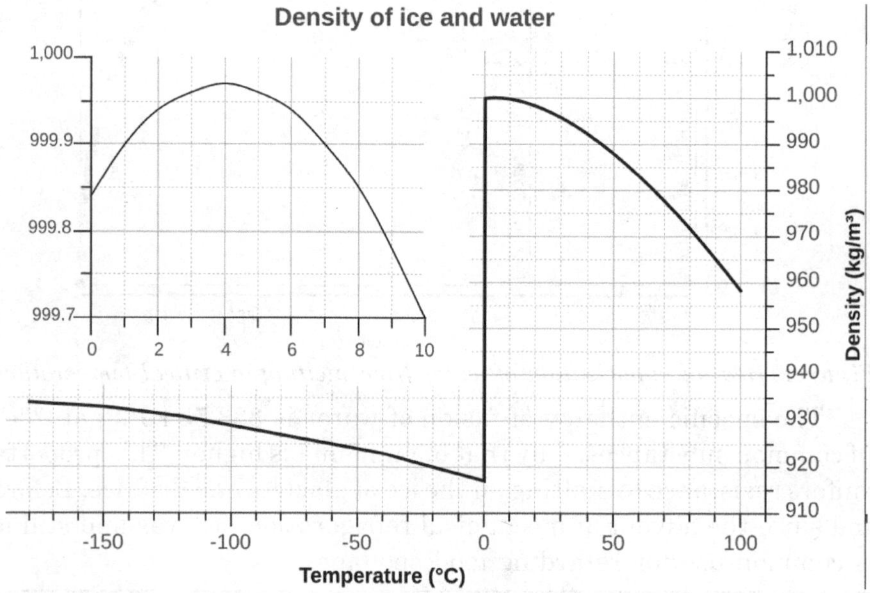

Figure: *Density of ice and water as a function of temperature*

Table: *The values below 0 °C refer to supercooled water.*

Density of Liquid Water	
Temp (°C)	*Density (kg/m³)*
+100	958.4
+80	971.8
+60	983.2
+40	992.2
+30	995.6502
+25	997.0479

Contd...

Temp (°C)	*Density (kg/m³)*
+22	997.7735
+20	998.2071
+15	999.1026
+10	999.7026
+4	999.9720
0	999.8395
–10	998.117
–20	993.547
–30	983.854

The density of water is approximately one gram per cubic centimeter. It is dependent on its temperature, but the relation is not linear and is unimodal rather than monotonic. When cooled from room temperature liquid water becomes increasingly dense, as with other substances, but at approximately 4 °C (39 °F), pure water reaches its maximum density. As it is cooled further, it expands to become less dense. This unusual negative thermal expansion is attributed to strong, orientation-dependent, intermolecular interactions and is also observed in molten silica.

The solid form of most substances is denser than the liquid phase; thus, a block of most solids will sink in the liquid. However, a block of ice floats in liquid water because ice is *less* dense. Upon freezing, the density of water decreases by about 9%. This is due to the 'cooling' of intermolecular vibrations allowing the molecules to form steady hydrogen bonds with their neighbours and thereby gradually locking into positions reminiscent of the hexagonal packing achieved upon freezing to ice I_h. Whereas the hydrogen bonds are shorter in the crystal than in the liquid, this locking effect reduces the average coordination number of molecules as the liquid approaches nucleation. Other substances that expand on freezing are silicon, gallium, germanium, antimony, bismuth, plutonium and also chemical compounds that form spacious crystal lattices with tetrahedral coordination.

Only ordinary hexagonal ice is less dense than the liquid. Under increasing pressure, ice undergoes a number of transitions to other allotropic forms with higher density than liquid water, such as ice II, ice III, high-density amorphous ice (HDA), and very-high-density amorphous ice (VHDA).

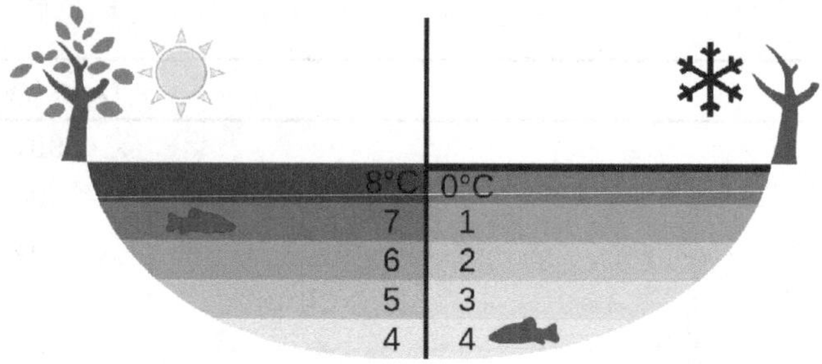

Figure: *Temperature distribution in a lake in summer and winter*

Water also expands significantly as the temperature increases. Water near the boiling point is about 96% as dense as water at 4 °C.

The melting point of ice is 0 °C (32 °F, 273.15 K) at standard pressure, however, pure liquid water can be supercooled well below that temperature without freezing if the liquid is not mechanically disturbed. It can remain in a fluid state down to its homogeneous nucleation point of approximately 231 K (–42 °C). The melting point of ordinary hexagonal ice falls slightly under moderately high pressures, but as ice transforms into its allotropes above 209.9 MPa (2,072 atm), the melting point increases markedly with pressure, i.e., reaching 355 K (82 °C) at 2.216 GPa (21,870 atm) (triple point of Ice VII).

A significant increase of pressure is required to lower the melting point of ordinary ice—the pressure exerted by an ice skater on the ice only reduces the melting point by approximately 0.09 °C (0.16 °F).

These properties of water have important consequences in its role in Earth's ecosystem. Water at a temperature of 4 °C will always accumulate at the bottom of freshwater lakes, irrespective of the temperature in the atmosphere. Since water and ice are poor conductors of heat (good insulators) it is unlikely that sufficiently deep lakes will freeze completely, unless stirred by strong currents that mix cooler and warmer water and accelerate the cooling. In warming weather, chunks of ice float, rather than sink to the bottom where they might melt extremely slowly. These properties therefore allow aquatic life in the lake to survive during the winter.

Density of Saltwater and Ice

The density of water is dependent on the dissolved salt content as well as the temperature of the water. Ice still floats in the oceans, otherwise they would freeze from the bottom up. However, the salt content of oceans lowers the freezing point by about 2 °C and lowers the temperature of the density maximum of water to the freezing point. This is why, in ocean water, the downward convection of colder water is *not* blocked by an expansion of water as it becomes colder near the freezing point. The oceans' cold water near the freezing point continues to sink. For this reason, any creature attempting to survive at the bottom of such cold water as the Arctic Ocean generally lives in water that is 4 °C colder than the temperature at the bottom of frozen-over fresh water lakes and rivers in the winter.

In cold countries, when the temperature of fresh water reaches 4 °C, the layers of water near the top in contact with cold air continue to lose heat energy and their temperature falls below 4 °C. On cooling below 4 °C, these layers do not sink but may rise up as fresh water has a maximum density at 4 °C. (Refer: Polarity and hydrogen bonding) Due to this, the layer of water at 4 °C remains at the bottom and above this layers of water 3 °C, 2 °C, 1 °C and 0 °C are formed. Because ice is a poor conductor of heat, it does not absorb heat energy from the water beneath the layer of ice which prevents the water freezing. Thus, aquatic creatures survive in such places.

As the surface of salt water begins to freeze (at –1.9 °C for normal salinity seawater, 3.5%) the ice that forms is essentially salt free with a density approximately equal to that of freshwater ice. This ice floats on the surface and the salt that is "frozen out" adds to the salinity and density of the seawater just below it, in a process known as *brine rejection*. This denser saltwater sinks by convection and the replacing seawater is subject to the same process. This provides essentially freshwater ice at –1.9 °C on the surface. The increased density of the seawater beneath the forming ice causes it to sink towards the bottom. On a large scale, the process of brine rejection and sinking cold salty water results in ocean currents forming to transport such water away from the Poles, leading to a global system of currents called the thermohaline circulation. One potential consequence of global warming is that the loss of Arctic and Antarctic ice could result in the loss of these currents as well, which could have unforeseeable consequences on near and distant climates.

Miscibility and Condensation

Water is miscible with many liquids, for example ethanol in all proportions, forming a single homogeneous liquid. On the other hand, water and most oils are *immiscible* usually forming layers according to increasing density from the top.

As a gas, water vapor is completely miscible with air. On the other hand the maximum water vapor pressure that is thermodynamically stable with the liquid (or solid) at a given temperature is relatively low compared with total atmospheric pressure. For example, if the vapor *partial pressure* is 2% of atmospheric pressure and the air is cooled from 25 °C, starting at about 22 °C water will start to condense, defining the dew point, and creating fog or dew.

The reverse process accounts for the fog *burning off* in the morning. If the humidity is increased at room temperature, for example, by running a hot shower or a bath, and the temperature stays about the same, the vapor soon reaches the pressure for phase change, and then condenses out as minute water droplets, commonly referred to as steam.

A gas in this context is referred to as *saturated* or 100% relative humidity, when the vapor pressure of water in the air is at the equilibrium with vapor pressure due to (liquid) water; water (or ice, if cool enough) will fail to lose mass through evaporation when exposed to saturated air. Because the amount of water vapor in air is small, *relative humidity*, the ratio of the partial pressure due to the water vapor to the saturated partial vapor pressure, is much more useful. Water vapor pressure above 100% relative humidity is called *supersaturated* and can occur if air is rapidly cooled, for example, by rising suddenly in an updraft.

Compressibility

The compressibility of water is a function of pressure and temperature. At 0 °C, at the limit of zero pressure, the compressibility is $5.1 \times 10^{-10}\ Pa^{-1}$. At the zero-pressure limit, the compressibility reaches a minimum of $4.4 \times 10^{-10}\ Pa^{-1}$ around 45 °C before increasing again with increasing temperature. As the pressure is increased, the compressibility decreases, being $3.9 \times 10^{-10}\ Pa^{-1}$ at 0 °C and 100 MPa.

The bulk modulus of water is 2.2 GPa. The low compressibility of non-gases, and of water in particular, leads to their often being assumed as incompressible. The low compressibility of water means that even in the deep oceans at 4 km depth, where pressures are 40 MPa, there is only a 1.8% decrease in volume.

Triple Point

The temperature and pressure at which solid, liquid, and gaseous water coexist in equilibrium is called the triple point of water. This point is used to define the units of temperature (the kelvin, the SI unit of thermodynamic temperature and, indirectly, the degree Celsius and even the degree Fahrenheit).

The various triple points of water		
Phases in stable equilibrium	***Pressure***	***Temperature***
liquid water, ice I_h, and water vapor	611.73 Pa	273.16 K (0.01 °C)
liquid water, ice I_h, and ice III	209.9 MPa	251 K (–22 °C)
liquid water, ice III, and ice V	350.1 MPa	–17.0 °C
liquid water, ice V, and ice VI	632.4 MPa	0.16 °C
ice I_h, Ice II, and ice III	213 MPa	–35 °C
ice II, ice III, and ice V	344 MPa	–24 °C
ice II, ice V, and ice VI	626 MPa	–70 °C

As a consequence, water's triple point temperature is a prescribed value rather than a measured quantity.

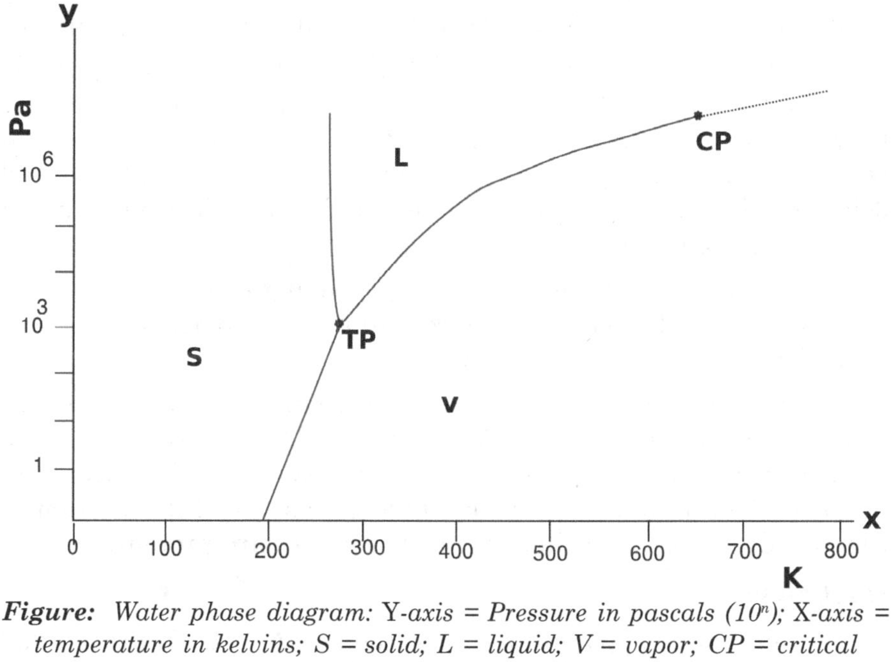

Figure: *Water phase diagram:* Y-*axis = Pressure in pascals (10^n);* X-*axis = temperature in kelvins; S = solid; L = liquid; V = vapor; CP = critical point; TP = triple point of water*

The triple point is at a temperature of 273.16 K (0.01 °C) by convention, and at a pressure of 611.73 Pa. This pressure is quite low, about $^1/_{166}$ of the normal sea level barometric pressure of 101,325 Pa. The atmospheric surface pressure on planet Mars is 610.5 Pa, which is remarkably close to the triple point pressure. The altitude of this surface pressure was used to define zero-elevation or "sea level" on that planet. Although it is commonly named as "*the* triple point of water", the stable combination of liquid water, ice I, and water vapor is but one of several triple points on the phase diagram of water. Gustav Heinrich Johann Apollon Tammann in Göttingen produced data on several other triple points in the early 20th century. Kamb and others documented further triple points in the 1960s.

Electrical Properties

Electrical Conductivity: Pure water containing no exogenous ions is an excellent insulator, but not even "deionized" water is completely free of ions. Water undergoes auto-ionization in the liquid state, when two water molecules form one hydroxide anion $\left(OH^-\right)$ and one hydronium cation (H_3O+).

Because water is such a good solvent, it almost always has some solute dissolved in it, often a salt. If water has even a tiny amount of such an impurity, then it can conduct electricity far more readily.

It is known that the theoretical maximum electrical resistivity for water is approximately 182 kΩ·m at 25 °C. This figure agrees well with what is typically seen on reverse osmosis, ultra-filtered and deionized ultra-pure water systems used, for instance, in semiconductor manufacturing plants.

A salt or acid contaminant level exceeding even 100 parts per trillion (ppt) in otherwise ultra-pure water begins to noticeably lower its resistivity by up to several kΩ·m.

In pure water, sensitive equipment can detect a very slight electrical conductivity of 0.055 μS/cm at 25 °C. Water can also be electrolyzed into oxygen and hydrogen gases but in the absence of dissolved ions this is a very slow process, as very little current is conducted. In ice, the primary charge carriers are protons.

Electrolysis

Water can be split into its constituent elements, hydrogen and oxygen, by passing an electric current through it. This process is called electrolysis. Water molecules naturally dissociate into H^+ and OH^- ions, which are attracted toward the cathode and anode, respectively. At the cathode, two H^+ ions pick up electrons and form gas. At the anode, four OH^- ions combine and release O_2 gas, molecular water, and four electrons.

The gases produced bubble to the surface, where they can be collected. The standard potential of the water electrolysis cell is 1.23 V at 25 °C.

Static Dielectric Constant

dielectric constant of water											
temperature /°C	0	10	20	30	40	50	60	70	80	90	100
ε	87.9	83.95	80.18	76.58	73.18	69.88	66.76	63.78	60.93	58.2	55.58

One of the important properties of water is that it has a high dielectric constant. This constant shows its ability to make electrostatic bonds with other molecules, meaning it can eliminate the attraction of the opposite charges of the surrounding ions.

Polarity and Hydrogen Bonding

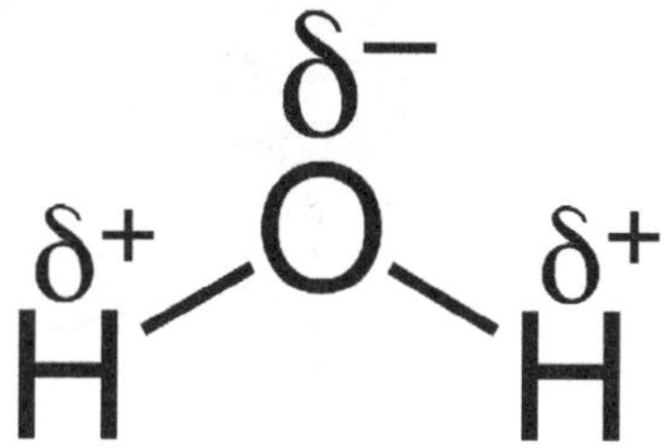

Figure: *A diagram showing the partial charges on the atoms in a water molecule*

An important feature of water is its polar nature. The water molecule forms an angle, with hydrogen atoms at the tips and oxygen at the vertex. This angle formed is 104.3 degrees as opposed to the typical tetrahedral angle of 109 degrees. Because oxygen has a higher electronegativity than hydrogen, the side of the molecule with the oxygen atom has a partial negative charge.

Also the presence of the lone pairs tend to push the oxygen away. An object with such a charge difference is called a dipole meaning two poles. The oxygen end is partially negative and the hydrogen end is partially positive, because of this the direction of the dipole moment points from the oxygen towards the centre of the hydrogens. The charge differences cause water molecules to be attracted to each other (the relatively positive areas being attracted to the relatively negative areas) and to other polar molecules. This attraction contributes to hydrogen bonding, and explains many of the properties of water, such as solvent action.

A water molecule can form a maximum of four hydrogen bonds because it can accept two and donate two hydrogen atoms. Other molecules like hydrogen fluoride, ammonia, methanol form hydrogen bonds but they do not show anomalous behaviour of thermodynamic, kinetic or structural properties like those observed in water. The answer to the apparent difference between water and other hydrogen bonding liquids lies in the fact that apart from water none of the hydrogen bonding molecules can form four hydrogen bonds, either due to an inability to donate/accept hydrogens or due to steric effects in bulky residues.

In water, local tetrahedral order due to the four hydrogen bonds gives rise to an open structure and a 3-dimensional bonding network, resulting in the anomalous decrease of density when cooled below 4 °C.

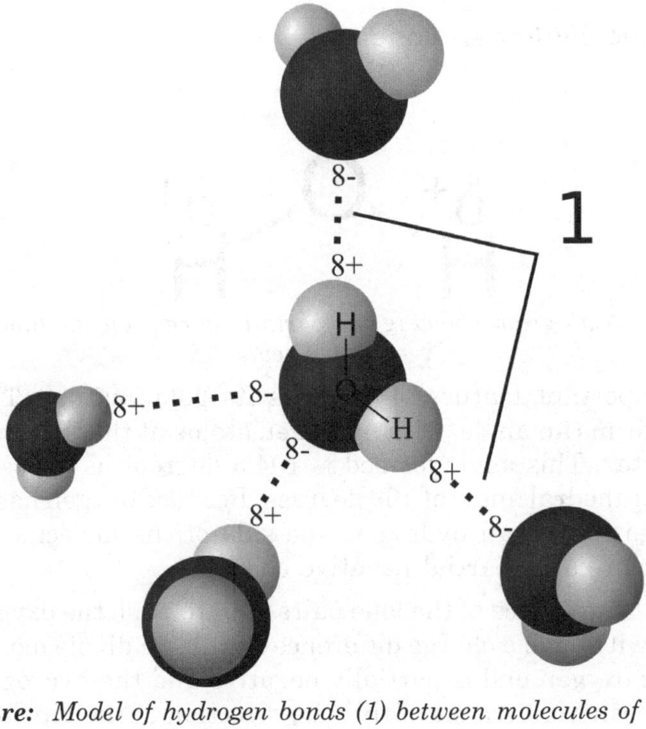

Figure: *Model of hydrogen bonds (1) between molecules of water*

Although hydrogen bonding is a relatively weak attraction compared to the covalent bonds within the water molecule itself, it is responsible for a number of water's physical properties. One such property is its relatively high melting and boiling point temperatures; more energy is required to break the hydrogen bonds between molecules.

The similar compound hydrogen sulfide (H_2S), which has much weaker hydrogen bonding, is a gas at room temperature even though it has twice the molecular mass of water. The extra bonding between water molecules also gives liquid water a large specific heat capacity. This high heat capacity makes water a good heat storage medium (coolant) and heat shield.

Cohesion and Adhesion

Water molecules stay close to each other (cohesion), due to the collective action of hydrogen bonds between water molecules. These hydrogen bonds are constantly breaking, with new bonds being formed with different water molecules; but at any given time in a sample of liquid water, a large portion of the molecules are held together by such bonds.

Water also has high adhesion properties because of its polar nature. On extremely clean/smooth glass the water may form a thin film because the molecular forces between glass and water molecules (adhesive forces) are stronger than the cohesive forces. In biological cells and organelles, water is in contact with membrane and protein surfaces that are hydrophilic; that is, surfaces that have a strong attraction to water. Irving Langmuir observed a strong repulsive force between hydrophilic surfaces. To dehydrate hydrophilic surfaces—to remove the strongly held layers of water of hydration—requires doing substantial work against these forces, called hydration forces. These forces are very large but decrease rapidly over a nanometer or less. They are important in biology, particularly when cells are dehydrated by exposure to dry atmospheres or to extracellular freezing.

Surface Tension

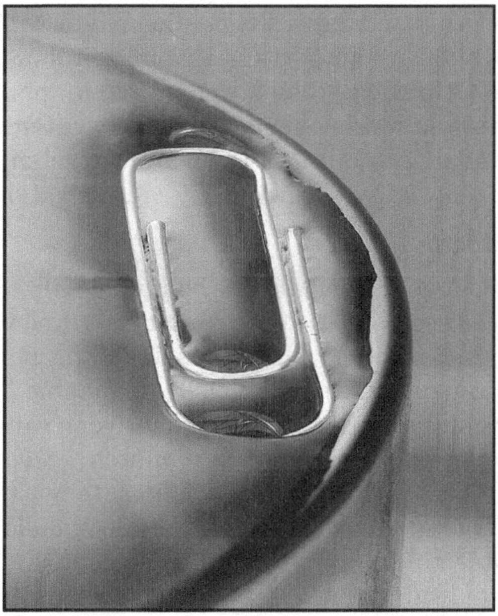

Figure: *This paper clip is under the water level, which has risen gently and smoothly. Surface tension prevents the clip from submerging and the water from overflowing the glass edges.*

Water has a high surface tension of 72.8 mN/m at room temperature, caused by the strong cohesion between water molecules, the highest of the common non-ionic, non-metallic liquids. This can be seen when small quantities of water are placed onto a sorption-free (non-adsorbent and non-absorbent) surface, such as polyethylene or Teflon, and the water stays together as drops. Just as significantly, air trapped in surface disturbances forms bubbles, which sometimes last long enough to transfer gas molecules to the water.

Another surface tension effect is capillary waves, which are the surface ripples that form around the impacts of drops on water surfaces, and sometimes occur with strong subsurface currents flowing to the water surface. The apparent elasticity caused by surface tension drives the waves.

Capillary Action

Due to an interplay of the forces of adhesion and surface tension, water exhibits capillary action whereby water rises into a narrow tube against the force of gravity. Water adheres to the inside wall of the tube and surface tension tends to straighten the surface causing a surface rise and more water is pulled up through cohesion. The process continues as the water flows up the tube until there is enough water such that gravity balances the adhesive force. Surface tension and capillary action are important in biology. For example, when water is carried through xylem up stems in plants, the strong intermolecular attractions (cohesion) hold the water column together and adhesive properties maintain the water attachment to the xylem and prevent tension rupture caused by transpiration pull.

Water as a Solvent

Water is also a good solvent, due to its polarity. Substances that will mix well and dissolve in water (e.g. salts) are known as hydrophilic ("water-loving") substances, while those that do not mix well with water (e.g. fats and oils), are known as hydrophobic ("water-fearing") substances. The ability of a substance to dissolve in water is determined by whether or not the substance can match or better the strong attractive forces that water molecules generate between other water molecules. If a substance has properties that do not allow it to overcome these strong intermolecular forces, the molecules are "pushed out" from the water, and do not dissolve. Contrary to the common misconception, water and hydrophobic substances do not "repel", and the hydration of a hydrophobic surface is energetically, but not entropically, favourable.

When an ionic or polar compound enters water, it is surrounded by water molecules (Hydration). The relatively small size of water molecules typically allows many water molecules to surround one molecule of solute. The partially negative dipole ends of the water are attracted to positively charged components of the solute, and vice versa for the positive dipole ends.

In general, ionic and polar substances such as acids, alcohols, and salts are relatively soluble in water, and non-polar substances such as fats and oils are not. Non-polar molecules stay together in water because it is energetically more favourable for the water molecules to hydrogen bond to each other than to engage in van der Waals interactions with non-polar molecules.

An example of an ionic solute is table salt; the sodium chloride, NaCl, separates into Na^+ cations and Cl^- anions, each being surrounded by water molecules. The ions are then easily transported away from their crystalline lattice into solution. An example of a nonionic solute is table sugar. The water dipoles make hydrogen bonds with the polar regions of the sugar molecule (OH groups) and allow it to be carried away into solution.

Water in Acid-base Reactions

Chemically, water is amphoteric: it can act as either an acid or a base in chemical reactions. According to the Brønsted-Lowry definition, an acid is defined as a species which donates a proton (a H^+ ion) in a reaction, and a base as one which receives a proton. When reacting with a stronger acid, water acts as a base; when reacting with a stronger base, it acts as an acid. For instance, water receives an H^+ ion from HCl when hydrochloric acid is formed:

$$\text{HCl}\left(\text{acid}\right) + \text{H}_2\text{O}\left(\text{base}\right) \rightleftharpoons \text{H}_3\text{O}^+ + \text{Cl}^-$$

In the reaction with ammonia, NH_3, water donates a H^+ ion, and is thus acting as an acid:

$$\text{NH}_3\left(\text{base}\right) + \text{H}_2\text{O}\left(\text{acid}\right) \rightleftharpoons \text{NH}_4^+ + \text{OH}^-$$

Because the oxygen atom in water has two lone pairs, water often acts as a Lewis base, or electron pair donor, in reactions with Lewis acids, although it can also react with Lewis bases, forming hydrogen bonds between the electron pair donors and the hydrogen atoms of water. HSAB theory describes water as both a weak hard acid and a weak hard base, meaning that it reacts preferentially with other hard species:

$$H^+ \text{(Lewis acid)} + H_2O\text{(Lewis base)} \rightarrow H3O^+$$

$$Fe^{3+} \text{(Lewis acid)} + H_2O\text{(Lewis base)} \rightarrow Fe(H_2O)_6^{3+}$$

$$Cl^- \text{(Lewis base)} + H_2O\text{(Lewis acid)} \rightarrow Cl(H_2O)_6^-$$

When a salt of a weak acid or of a weak base is dissolved in water, water can partially hydrolyze the salt, producing the corresponding base or acid, which gives aqueous solutions of soap and baking soda their basic pH:

$$Na_2CO_3 + H_2O \rightleftharpoons NaOH + NaHCO_3$$

Ligand Chemistry

Water's Lewis base character makes it a common ligand in transition metal complexes, examples of which range from solvated ions, such as $Fe(H_2O)_6^{3+}$, to perrhenic acid, which contains two water molecules coordinated to a rhenium atom, to various solid hydrates, such as $CoCl_2 \cdot 6H_2O$. Water is typically a monodentate ligand, it forms only one bond with the central atom.

Organic Chemistry

As a hard base, water reacts readily with organic carbocations, for example in hydration reaction, in which a hydroxyl group (OH^-) and an acidic proton are added to the two carbon atoms bonded together in the carbon-carbon double bond, resulting in an alcohol. When addition of water to an organic molecule cleaves the molecule in two, hydrolysis is said to occur. Notable examples of hydrolysis are saponification of fats and digestion of proteins and polysaccharides. Water can also be a leaving group in S_N2 substitution and E2 elimination reactions, the latter is then known as dehydration reaction.

Acidity in Nature

Pure water has the concentration of hydroxide ions (OH^-) equal to that of the hydronium (H_3O^+) or hydrogen (H^+) ions, which gives pH of 7 at 298 K. In practice, pure water is very difficult to produce. Water left exposed to air for any length of time will dissolve carbon dioxide, forming a dilute solution of carbonic acid, with a limiting pH of about 5.7. As cloud droplets form in the atmosphere and as raindrops fall through the air minor amounts of CO_2 are absorbed, and thus most rain is slightly acidic. If high amounts of nitrogen and sulphur oxides are present in the air, they too will dissolve into the cloud and rain drops, producing acid rain.

Water in Redox Reactions

Water contains hydrogen in oxidation state +1 and oxygen in oxidation state –2. Because of that, water oxidizes chemicals with reduction potential below the potential of H^+/H_2, such as hydrides, alkali and alkaline earth metals (except for beryllium), etc. Some other reactive metals, such as aluminium, are oxidized by water as well, but their oxides are not soluble, and the reaction stops because of passivation. Note, however, that rusting of iron is a reaction between iron and oxygen, dissolved in water, not between iron and water.

$$2\ Na + 2\ H_2O \rightarrow 2\ NaOH + H_2$$

Water itself can be oxidized, emitting oxygen gas, but very few oxidants react with water even if their reduction potential is greater than the potential of $O_2 / O2^-$

- Almost all such reactions require a catalyst.

$$4\ Ag\ F_2 + 2\ H_2O \rightarrow 4\ AgF + 4\ HF + O_2$$

Geochemistry

Action of water on rock over long periods of time typically leads to weathering and water erosion, physical processes that convert solid rocks and minerals into soil and sediment, but under some conditions chemical reactions with water occur as well, resulting in metasomatism or mineral hydration, a type of chemical alteration of a rock which produces clay minerals in nature and also occurs when Portland cement hardens.

Water ice can form clathrate compounds, known as clathrate hydrates, with a variety of small molecules that can be embedded in its spacious crystal lattice. The most notable of these is methane clathrate, $4CH_4 \cdot 23H_2O$, naturally found in large quantities on the ocean floor.

Transparency

Water is relatively transparent to visible light, near ultraviolet light, and far-red light, but it absorbs most ultraviolet light, infrared light, and microwaves. Most photoreceptors and photosynthetic pigments utilize the portion of the light spectrum that is transmitted well through water. Microwave ovens take advantage of water's opacity to microwave radiation to heat the water inside of foods. The very weak onset of absorption in the red end of the visible spectrum lends water its intrinsic blue hue.

Heavy Water and Isotopologues

Several isotopes of both hydrogen and oxygen exist, giving rise to several known isotopologues of water. Hydrogen occurs naturally in three isotopes. The most common (^{1}H) accounting for more than 99.98% of hydrogen in water, consists of only a single proton in its nucleus. A second, stable isotope, deuterium (chemical symbol *D* or ^{2}H), has an additional neutron. Deuterium oxide, D_2O, is also known as heavy water because of its higher density. It is used in nuclear reactors as a neutron moderator. The third isotope, tritium, has 1 proton and 2 neutrons, and is radioactive, decaying with a half-life of 4500 days. T_2O exists in nature only in minute quantities, being produced primarily via cosmic ray-induced nuclear reactions in the atmosphere. Water with one deuterium atom HDO occurs naturally in ordinary water in low concentrations (~0.03%) and D_2O in far lower amounts (0.000003%).

The most notable physical differences between H_2O and D_2O, other than the simple difference in specific mass, involve properties that are affected by hydrogen bonding, such as freezing and boiling, and other kinetic effects. The difference in boiling points allows the isotopologues to be separated.

Consumption of pure isolated D_2O may affect biochemical processes – ingestion of large amounts impairs kidney and central nervous system function. Small quantities can be consumed without any ill-effects, and even very large amounts of heavy water must be consumed for any toxicity to become apparent.

Oxygen also has three stable isotopes, with 16O present in 99.76%, 17O in 0.04%, and 18 O in 0.2% of water molecules.

History

The first decomposition of water into hydrogen and oxygen, by electrolysis, was done in 1800 by an English chemist William Nicholson. In 1805, Joseph Louis Gay-Lussac and Alexander von Humboldt showed that water is composed of two parts hydrogen and one part oxygen.

Gilbert Newton Lewis isolated the first sample of pure heavy water in 1933.

The properties of water have historically been used to define various temperature scales. Notably, the Kelvin, Celsius, Rankine, and Fahrenheit scales were, or currently are, defined by the freezing and boiling points of water. The less common scales of Delisle, Newton, Réaumur and Rømer were defined similarly. The triple point of water is a more commonly used standard point today.

Systematic Naming

The accepted IUPAC name of water is *oxidane* or simply *water*, or its equivalent in different languages, although there are other systematic names which can be used to describe the molecule.

The simplest systematic name of water is *hydrogen oxide.* This is analogous to related compounds such as hydrogen peroxide, hydrogen sulfide, and deuterium oxide (heavy water). Another systematic name, *oxidane*, is accepted by IUPAC as a parent name for the systematic naming of oxygen-based substituent groups, although even these commonly have other recommended names. For example, the name hydroxyl is recommended over *oxidanyl* for the –OH group. The name oxane is explicitly mentioned by the IUPAC as being unsuitable for this purpose, since it is already the name of a cyclic ether also known as tetrahydropyran.

The polarized form of the water molecule, H^+OH^-, is also called hydron hydroxide by IUPAC nomenclature.

Dihydrogen monoxide (DHMO) is a rarely used name of water. This term has been used in various hoaxes that call for this "lethal chemical" to be banned, such as in the dihydrogen monoxide hoax. Other systematic names for water include *hydroxic acid*, *hydroxylic acid*, and *hydrogen hydroxide.* Both acid and alkali names exist for water because it is amphoteric (able to react both as an acid or an alkali). None of these exotic names are used widely.

Rock Properties

Chemical Characteristics

The more common rock constituents are nearly all oxides; chlorides, sulphides and fluorides are the only important exceptions to this and their total amount in any rock is usually much less than 1%. F. W. Clarke has calculated that a little more than 47% of the Earth's crust consists of oxygen. It occurs principally in combination as oxides, of which the chief are silica, alumina, iron oxides, and various carbonates (calcium carbonate, magnesium carbonate, sodium carbonate, and potassium carbonate). The silica functions principally as an acid, forming silicates, and all the commonest minerals of igneous rocks are of this nature. From a computation based on 1672 analyses of numerous kinds of rocks Clarke arrived at the following as the average percentage composition: SiO_2=59.71, Al_2O_3=15.41, Fe_2O_3=2.63, FeO=3.52, MgO=4.36, CaO=4.90, Na_2O=3.55, K_2O=2.80, H_2O=1.52,

TiO_2=0.60, P_2O_5=0.22, total 99.22%). All the other constituents occur only in very small quantities, usually much less than 1%.

These oxides combine in a haphazard way. For example, potash (potassium carbonate) and soda (sodium carbonate) combine to produce feldspars. In some cases they may take other forms, such as nepheline, leucite, and muscovite, but in the great majority of instances they are found as feldspar. Phosphoric acid with lime (calcium carbonate) forms apatite. Titanium dioxide with ferrous oxide gives rise to ilmenite. Part of the lime forms lime feldspar.

Magnesium carbonate and iron oxides with silica crystallize as olivine or enstatite, or with alumina and lime form the complex ferro-magnesian silicates of which the pyroxenes, amphiboles, and biotites are the chief. Any excess of silica above what is required to neutralize the bases will separate out as quartz; excess of alumina crystallizes as corundum. These must be regarded only as general tendencies. It is possible, by rock analysis, to say approximately what minerals the rock contains, but there are numerous exceptions to any rule.

Mineral Constitution

Hence we may say that except in acid or siliceous rocks containing 66% of silica and over, quartz will not be abundant. In basic rocks (containing 20% of silica or less) it is rare and accidental. If magnesia and iron be above the average while silica is low, olivine may be expected; where silica is present in greater quantity over ferro-magnesian minerals, such as augite, hornblende, enstatite or biotite, occur rather than olivine. Unless potash is high and silica relatively low, leucite will not be present, for leucite does not occur with free quartz. Nepheline, likewise, is usually found in rocks with much soda and comparatively little silica. With high alkalis, soda-bearing pyroxenes and amphiboles may be present. The lower the percentage of silica and the alkalis, the greater is the prevalence of calcium feldspar as contracted with soda or potash feldspar. Clarke has calculated the relative abundance of the principal rock-forming minerals with the following results: apatite=0.6, titanium minerals=1.5, quartz=12.0, feldspars=59.5, biotite=3.8, hornblende and pyroxene=16.8, total=94.2%. This, however, can only be a rough approximation.

The other determining factor, namely the physical conditions attending consolidation, plays on the whole a smaller part, yet is by no means negligible, as a few instances will prove. Certain minerals are practically confined to deep-seated intrusive rocks, e.g., microcline, muscovite, diallage. Leucite is very rare in plutonic masses; many minerals have special peculiarities in microscopic character according to whether they crystallized in depth or near the surface, e.g., hypersthene, orthoclase, quartz.

There are some curious instances of rocks having the same chemical composition, but consisting of entirely different minerals, e.g., the hornblendite of Gran, in Norway, which contains only hornblende, has the same composition as some of the camptonites of the same locality that contain feldspar and hornblende of a different variety. In this connection we may repeat what has been said above about the corrosion of porphyritic minerals in igneous rocks.

In rhyolites and trachytes, early crystals of hornblende and biotite may be found in great numbers partially converted into augite and magnetite. Hornblende and biotite were stable under the pressures and other conditions below the surface, but unstable at higher levels. In the ground-mass of these rocks, augite is almost universally present. But the plutonic representatives of the same magma, granite and syenite contain biotite and hornblende far more commonly than augite.

Felsic, Intermediate and Mafic Igneous Rocks

Those rocks that contain the most silica, and on crystallizing yield free quartz, form a group generally designated the "felsic" rocks. Those again that contain least silica and most magnesia and iron, so that quartz is absent while olivine is usually abundant, form the "mafic" group. The "intermediate" rocks include those characterized by the general absence of both quartz and olivine. An important subdivision of these contains a very high percentage of alkalis, especially soda, and consequently has minerals such as nepheline and leucite not common in other rocks. It is often separated from the others as the "alkali" or "soda" rocks, and there is a corresponding series of mafic rocks. Lastly a small sub-group rich in olivine and without feldspar has been called the "ultramafic" rocks. They have very low percentages of silica but much iron and magnesia.

Except these last, practically all rocks contain felspars or feldspathoid minerals. In the acid rocks the common feldspars are orthoclase, perthite, microcline, and oligoclase—all having much silica and alkalis. In the mafic rocks labradorite, anorthite and bytownite prevail, being rich in lime and poor in silica, potash and soda. Augite is the most common ferro-magnesian in mafic rocks, but biotite and hornblende are on the whole more frequent in felsic rocks.

Most Common Minerals	*Acid*	*Intermediate*		*Mafic*	*Ultramafic*
	Quartz Orthoclase (and Oligoclase), Mica, Hornblende, Augite	*Little or no Quartz: Orthoclase hornblende, Augite, Biotite*	*Little or no Quartz: Plagioclase Hornblende, Augite, Biotite*	*No Quartz Plagioclase Augite, Olivine*	*No Felspar Augite, Hornblende, Olivine*
Plutonic or Abyssal type	Granite	Syenite	Diorite	Gabbro	Peridotite
Intrusive or Hypabyssal type	Quartz-porphyry	Orthoclase-porphyry	Porphyrite	Dolerite	Picrite
Lavas or Effusive type	Rhyolite, Obsidian	Trachyte	Andesite	Basalt	Limburgite

Rocks that contain leucite or nepheline, either partly or a wholly replacing felspar, are not included in this table. They are essentially of intermediate or of mafic character. We might in consequence regard them as varieties of syenite, diorite, gabbro, etc., in which feldspathoid minerals occur, and indeed there are many transitions between syenites of ordinary type and nepheline — or leucite — syenite, and between gabbro or dolerite and theralite or essexite.

But, as many minerals develop in these "alkali" rocks that are uncommon elsewhere, it is convenient in a purely formal classification like that outlined here to treat the whole assemblage as a distinct series.

	Nepheline and Leucite-bearing Rocks		
Most Common Minerals	*Alkali Feldspar, Nepheline or Leucite, Augite, Hornblend, Biotite*	*Soda Lime Feldspar, Nepheline or Leucite, Augite, Hornblende (Olivine)*	*Nepheline or Leucite, Augite, Hornblende, Olivine*
Plutonic type	Nepheline-syenite, Leucite-syenite, Nepheline-porphyry	Essexite and Theralite	Ijolite and Missourite
Effusive type or Lavas	Phonolite, Leucitophyre	Tephrite and Basanite	Nepheline-basalt, Leucite-basalt

This classification is based essentially on the mineralogical constitution of the igneous rocks. Any chemical distinctions between the different groups, though implied, are relegated to a subordinate position. It is admittedly artificial but it has grown up with the growth of the science and is still adopted as the basis on which more minute subdivisions are erected.

The subdivisions are by no means of equal value. The syenites, for example, and the peridotites, are far less important than the granites, diorites and gabbros. Moreover, the effusive andesites do not always correspond to the plutonic diorites but partly also to the gabbros.

As the different kinds of rock, regarded as aggregates of minerals, pass gradually into one another, transitional types are very common and are often so important as to receive special names. The quartz-syenites and nordmarkites may be interposed between granite and syenite, the tonalites and adamellites between granite and diorite, the monzoaites between syenite and diorite, norites and hyperites between diorite and gabbro, and so on.

Formation of Rocks

Figure: *Stone*

The three main ways rocks are formed:

- Sedimentary rocks are formed through the gradual accumulation of sediment: for example, sand on a beach or mud on a river bed. As the sediment is buried it is compacted as more and more material is deposited on top. Eventually the sediment will become so dense that it is essentially rock. This process is known as lithification.
- Igneous rocks are rocks which have crystallized from a melt or magma. The melt is made up of various components of pre-existing rocks which have been subjected to melting either at

subduction zones or within the Earth's mantle. The melt is hot and so passes upward through cooler country rock. As it moves it cools and various rock types will form through a process known as fractional crystallization. Igneous rocks can be seen at mid ocean ridges, areas of island arc volcanism or in intra-plate hotspots.

- Metamorphic rocks are rocks which once existed as igneous or sedimentary rocks but have been subjected to varying degrees of pressure and heat within the Earth's crust. The processes involved will change the composition and fabric of the rock and their original nature is often hard to distinguish. Metamorphic rocks are typically found in areas of mountain building.

Rock Synthesis

The synthetic investigation of rocks proceeds by experimental work that attempts to reproduce different rock types and to elucidate their origins and structures. In many cases no experiment is necessary. Every stage in the origin of clays, sands and gravels can be seen in process around us, but where these have been converted into coherent shales, sandstone and conglomerates, and still more where they have experienced some degree of metamorphism, there are many obscure points about their history upon which experiment may yet throw light. Attempts have been made to reproduce igneous rocks, by fusion of mixtures of crushed minerals or of chemicals in specially contrived furnaces.

The earliest researches of this sort are those of Faujas St Fond and of de Saussure, but Sir James Hall really laid the foundations of this branch of petrology. He showed (1798) that the whinstones (diabases) of Edinburgh were fusible and if rapidly cooled yielded black vitreous masses closely resembling natural pitchstones and obsidians, if cooled more slowly they consolidated as crystalline rocks not unlike the whinstones themselves and containing olivine, augite and feldspar (the essential minerals of these rocks).

Many years later Daubrée, Delesse and others carried on similar experiments, but the first notable advance was made in 1878, when Fouqué and Lévy began their researches. They succeeded in producing such rocks as porphyrite, leucite-tephrite, basalt and dolerite, and obtained also various structural modifications well known in igneous rocks, e.g. the porphyritic and the ophitic. Incidentally they showed that while many basic rocks (basalts, etc.) could be perfectly imitated in the laboratory, the acid rocks could not, and advanced the explanation that for the crystallization of the latter the gases never absent in natural rock magmas were indispensable mineralizing agents. It has subsequently been proved that steam, or such volatile substances as certain borates, molybdates, chlorides, fluorides, assist in the formation of orthoclase, quartz and mica (the minerals of granite). Sir James Hall also made the first contribution to the experimental study of metamorphic rocks by converting chalk into marble by heating it in a closed gun-barrel, which prevented the escape of the carbonic acid at high temperatures.

In 1901 Adams and Nicholson carried this a stage further by subjecting marble to great pressures in hydraulic presses and have shown how the foliated structures, frequent in natural marbles, may be produced artificially.

Chapter 5
Tubular Technology

Expandable tubular technology is a system for increasing the diameter of the casing or liner of an oil well by up to 30% after it has been run down-hole.

Challenges

Two challenges facing the Oil & Gas industry are accessing new reservoirs that currently cannot be reached economically and maintaining profitable production from producing older fields. Expandable tubular technology is considered one of the most revolutionary technologies that has benefited the oil and gas industry during the last 10 years. The development of this technology was initiated by the industry need to reduce drilling costs, increase production of tubing constrained wells and to enable operators to access reservoirs that could otherwise not be reached economically. Expanded casing applications concentrate on reducing the telescopic profile of well designs through a downhole tube expansion process. Wider applications of the technology exist for example water shut off, and casing repairs in old wells.

Telescoping Well Design

To understand the enormous potential value of expandable tubular technology one must first understand the concept of telescoping well design and the constraints it places on conventional drilling.

During the planning stages, one of the primary considerations that the well engineer takes into account is the density (or "weight") of the drilling fluid. It must be heavy enough to suppress any pressured fluids contained by the rock (thus avoiding a blowout), and light enough to avoid breaking down the rock itself. Both parameters generally increase with depth, and create a window between which the drilling fluid density is safe. However, the mud density required to suppress fluid pressures at 9000' would usually break down the formation further up the well at 1000'. Therefore, the well is drilled in sections by running casing strings to cover depth ranges between which the required mud densities are suitable for the entire range. However, each bit must be smaller than the previous casing string, which in turn has to be smaller than the previous hole.

This requirement of well design creates a well where each hole section is ever-decreasing in diameter. The only conventional way to combat this effect is to start with an enormous hole at the top (sometimes 30") in order to run five casing strings and still end up with a 6" hole in the targeted reservoir. And this well design leaves little room for error: if the drillers are forced to run a casing string on account of unexpected formation pressures then the target formation may be impossible to reach. Thus there is an enormous carrot in a casing system that loses minimal - or possibly no - diameter while still isolating a whole section. Many current systems have the aim of producing a monobore well, that is, a well with a single diameter from top to bottom. But the shallow wells can the drilled with uniform diameter.

Technology

Figure: *Image of Expanded Tubular FE Analysis*

To reduce the loss of diameter each time a new casing string or liner is set, a cold working process has been developed whereby the casing or liner can be expanded by up to 20% in diameter after being run down-hole.

For this purpose, an expansion tool that exceeds the inner diameter of the tube by the required amount of expansion is forced through the pipe. This is done either hydraulically, by applying mud pressure, or mechanically, by pulling the conical/tapered expansion tool. The expansion needs to be reliable, when expanding several thousand feet below the surface. This can be from 30–6,000 ft in length.

Carey Naquin was responsible for the first commercial and successful application of expandable tubular technology.

Applications

The applications can be groups into two main categories – Cased hole and Open hole. Cased hole work is mainly down during the work over or completion phase of a well. The open hole products are used during the drilling period of a well.

The products developed and available today in cased hole work are the expandable liner hanger and the cased hole clad. The expandable liner hanger is basically an evolution of existing equipment currently used in the oil industry, a product with better thru bore and envisaged higher reliability. The Case hole clad provide a casing patch across a damaged section of casing, or to close off previously perforated casing. This product has two main advantages – minimal through bore loss [basically two times the wall thickness of tubular being expanded] and high pressure integrity performance.

Open hole applications is where expandable technology brings real advantages to the operator. Currently the products available are open hole liner and open hole clads.

Offshore Drilling

Figure: *An oil drilling platform off the coast of Santa Barbara, CA - 6 December, 2011*

Offshore drilling refers to a mechanical process where a wellbore is drilled through the seabed. It is typically carried out in order to explore for and subsequently extract petroleum which lies in rock formations beneath the seabed. Most commonly, the term is used to describe drilling activities on the continental shelf, though the term can also be applied to drilling in lakes, inshore waters and inland seas.

Offshore drilling presents environmental challenges, both from the produced hydrocarbons and the materials used during the drilling operation. Controversies include the ongoing US offshore drilling debate.

There are many different types of facilities from which offshore drilling operations take place. These include bottom founded drilling rigs (jackup barges and swamp barges), combined drilling and production facilities either bottom founded or floating platforms, and deepwater mobile offshore drilling units (MODU) including semi-submersibles and drillships. These are capable of operating in water depths up to 10,000 feet (3,000 m). In shallower waters the mobile units are anchored to the seabed, however in deeper water (>5,000 feet (1,500 m)) the semisubmersibles or drillships are maintained at the required drilling location using dynamic positioning.

History

Around 1891, the first submerged oil wells were drilled from platforms built on piles in the fresh waters of the Grand Lake St. Marys (a.k.a. Mercer County Reservoir) in Ohio. The wells were developed by small local companies such as Bryson, Riley Oil, German-American, and Banker's Oil.

Around 1896, the first submerged oil wells in salt water were drilled in the portion of the Summerland field extending under the Santa Barbara Channel in California. The wells were drilled from piers extending from land out into the channel.

Other notable early submerged drilling activities occurred on the Canadian side of Lake Erie in the 1900s and Caddo Lake in Louisiana in the 1910s. Shortly thereafter wells were drilled in tidal zones along the Texas and Louisiana gulf coast. The Goose Creek Oil Field near Baytown, Texas is one such example. In the 1920s drilling activities occurred from concrete platforms in Venezuela's Lake Maracaibo.

One of the oldest subsea wells is the Bibi Eibat well, which came on stream in 1923 in Azerbaijan. The well was located on an artificial island in a shallow portion of the Caspian Sea. In the early 1930s, the Texas Co., later Texaco (now Chevron) developed the first mobile steel barges for drilling in the brackish coastal areas of the Gulf of Mexico.

In 1937, Pure Oil (now Chevron) and its partner Superior Oil (now ExxonMobil) used a fixed platform to develop a field 1 mile (1.6 km) offshore of Calcasieu Parish, Louisiana in 14 feet (4.3 m) of water.

In 1946, Magnolia Petroleum (now ExxonMobil) drilled at a site 18 miles (29 km) off the coast, erecting a platform in 18 feet (5.5 m) of water off St. Mary Parish, Louisiana.

In early 1947, Superior Oil erected a drilling and production platform in 20 feet (6.1 m) of water some 18 miles (29 km) off Vermilion Parish, La. But it was Kerr-McGee Oil Industries (now Anadarko Petroleum), as operator for partners Phillips Petroleum (ConocoPhillips) and Stanolind Oil & Gas (BP) that completed its historic Ship Shoal Block 32 well in October 1947, months before Superior actually drilled a discovery from their Vermilion platform farther offshore. In any case, that made Kerr-McGee's well the first oil discovery drilled out of sight of land.

When offshore drilling moved into deeper waters of up to 100 feet (30 m), fixed platform rigs were built, until demands for drilling equipment was needed in the 100 feet (30 m) to 400 feet (120 m) depth of the Gulf of Mexico, the first jack-up rigs began appearing from specialized offshore drilling contractors such as forerunners of ENSCO International.

The first semi-submersible resulted from an unexpected observation in 1961. Blue Water Drilling Company owned and operated the four-column submersible Blue Water Rig No.1 in the Gulf of Mexico for Shell Oil Company. As the pontoons were not sufficiently buoyant to support the weight of the rig and its consumables, it was towed between locations at a draught mid-way between the top of the pontoons and the underside of the deck. It was noticed that the motions at this draught were very small, and Blue Water Drilling and Shell jointly decided to try operating the rig in the floating mode. The concept of an anchored, stable floating deep-sea platform had been designed and tested back in the 1920s by Edward Robert Armstrong for the purpose of operating aircraft with an invention known as the 'seadrome'. The first purpose-built drilling semi-submersible *Ocean Driller* was launched in 1963. Since then, many semi-submersibles have been purpose-designed for the drilling industry mobile offshore fleet. The first offshore drillship was the *CUSS 1* developed for the Mohole project to drill into the Earth's crust.

As of June, 2010, there were over 620 mobile offshore drilling rigs (Jackups, semisubs, drillships, barges) available for service in the competitive rig fleet.

One of the world's deepest hubs is currently the Perdido in the Gulf of Mexico, floating in 2,438 metres of water. It is operated by Royal Dutch Shell and was built at a cost of $3 billion. The deepest operational platform is the Petrobras America Cascade FPSO in the Walker Ridge 249 field in 2,600 metres of water.

Main Offshore Fields

Notable offshore fields today are found in the North Sea, the Gulf of Mexico (mostly offshore Texas and Louisiana, but also Mississippi and Alabama), California (in the Santa Barbara basin), the Caspian Sea (notably some major fields offshore Azerbaijan) the Campos and Santos Basins off the coasts of Brazil, Newfoundland and Nova Scotia (Atlantic Canada), several fields off West Africa most notably west of Nigeria and Angola, as well as offshore fields in South East Asia and Sakhalin, Russia. Also major offshore oil fields are located in the Persian Gulf such as Safaniya, Manifa, and Marjan which belong to Saudi Arabia and are developed by Saudi Aramco.

Challenges

Offshore oil and gas production is more challenging than land-based installations due to the remote and harsher environment. Much of the innovation in the offshore petroleum sector concerns overcoming these challenges, including the need to provide very large production facilities. Production and drilling facilities may be very large and a large investment, such as the Troll A platform standing on a depth of 300 metres.

Another type of offshore platform may float with a mooring system to maintain it on location. While a floating system may be lower cost in deeper waters than a fixed platform, the dynamic nature of the platforms introduces many challenges for the drilling and production facilities. The ocean can add several hundred metres or more to the fluid column. The addition increases the equivalent circulating density and downhole pressures in drilling wells, as well as the energy needed to lift produced fluids for separation on the platform.

The trend today is to conduct more of the production operations subsea, by separating water from oil and re-injecting it rather than pumping it up to a platform, or by flowing to onshore, with no installations visible above the sea. Subsea installations help to exploit resources at progressively deeper waters, locations which have been inaccessible, and overcome challenges posed by sea ice, such as in the Barents Sea. One such challenge in shallower environments is seabed gouging by drifting ice features (means of protecting offshore installations against ice action includes burial in the seabed).

Offshore manned facilities also present logistics and human resources challenges. An offshore oil platform is a small community in itself with cafeteria, sleeping quarters, management, and other support functions. In the North Sea, staff members are transported by helicopter for a two-week shift. They usually receive higher salary than onshore workers do. Supplies and waste are transported by ship, and the supply deliveries need to be carefully planned because storage space on the platform is limited. Today, much effort goes into relocating as many of the personnel as possible onshore, where management and technical experts are in touch with the platform by video conferencing. An onshore job is also more attractive for the aging workforce in the petroleum industry, at least in the western world. These efforts among others are contained in the established term integrated operations. The increased use of subsea facilities helps achieve the objective of keeping more workers onshore. Subsea facilities are also easier to expand, with new separators or different modules for different oil types, and are not limited by the fixed floor space of an above-water installation.

Effects on the Environment

Offshore oil production involves environmental risks, most notably oil spills from oil tankers or pipelines transporting oil from the platform to onshore facilities, and from leaks and accidents on the platform. Produced water is also generated, which is excess water from oil or gas production and includes varying amounts of oil, or other chemicals used in, or resulting from, oil production. According to the organization Culture Change, a Gulf of Mexico rig dumps about 90,000 tons of drilling fluid and cuttings over its lifetime, with its wells also contributing with heavy metals.

Deepwater Drilling

Deepwater drilling is the process of oil and gas exploration and production in depths of more than 500 feet. By this definition, there are approximately 600 deepwater wells in the Gulf of Mexico.

It has been economically infeasible for many years, but with rising oil prices, more companies are investing in this area. Major companies working in this sector include Diamond Offshore, TransOcean, and Baker Hughes.

There are basically two kinds of mobile deepwater drilling rigs: semi-submersible drilling rigs and drillships. Drilling can also be performed from a fixed-position installation such as a floating spar (platform) or tension-leg platform.

Recent industry analysis has estimated that the total expenditure in the global deepwater infrastructure market will reach $145bn in 2011. "Not all oil is accessible on land or in shallow water. You can find some oil deposits buried deep under the ocean floor." ...Using sonic equipment, oil companies determine the drilling sites most likely to produce oil. Then they use a mobile offshore drilling unit (MODU) to dig the initial well. Some units are converted into production rigs, meaning they switch from drilling for oil to capturing oil once it's found. Most of the time, the oil company will replace the MODU with a more permanent oil production rig to capture oil." ...The MODU's job is to drill down into the ocean's floor to find oil deposits. The part of the drill that extends below the deck and through the water is called the riser. The riser allows for drilling fluids to move between the floor and the rig. Engineers lower a drill string — a series of pipes designed to drill down to the oil deposit — through the riser."

"The expansion of deepwater drilling is happening despite accidents in offshore fields..." In 2010 a large explosion occurred killing 11 workers and spilling millions of gallons of oil into the Gulf of Mexico while a BP oil rig was drilling in deep waters. "...Despite the risks, the deepwater drilling trend is spreading in the Mediterranean and off the coast of East Africa after a string of huge discoveries of natural gas." "The reason for the resumption of such drilling, analyst say, is continuing high demand for energy worldwide." America needs the oil so it is almost necessary to face the high risks that come with deepwater drilling in order for Americans to continue to carry on their everyday lives. "The Energy Department recently projected that gulf oil production would expand from its 2011 level of 1.3 million barrels a day, still nearly a quarter of total domestic production, to 2 million barrels a day by 2020."

Shallow Water Drilling

Shallow water drilling is the process of oil and gas exploration and production in less than 500 feet of water. Shallow water drilling differs from deepwater drilling in several key aspects. Shallow water rigs have legs that reach the bottom of the sea floor and have blowout preventers (BOPs) above the surface of the water that are accessible for inspection, maintenance and repair, and can be controlled either remotely or manually in case of an emergency. Shallow water wells primarily produce natural gas, and are drilled in known areas and mature reservoirs.

Following the Deepwater Horizon oil spill, the U.S Department of Interior imposed a moratorium on all offshore drilling, both deepwater and shallow water. The ban on shallow water drilling was lifted in May 2010. However, new regulations imposed on shallow water drilling have slowed the issuance of permits for new shallow water wells. Shallow water operators have called this permit slowdown a "de facto moratorium" that has forced them to idle rig workers and decrease production.

Underbalanced Drilling

Underbalanced drilling, or UBD, is a procedure used to drill oil and gas wells where the pressure in the wellbore is kept lower than the fluid pressure in the formation being drilled. As the well is being drilled, formation fluid flows into the wellbore and up to the surface. This is the opposite of the usual situation, where the wellbore is kept at a pressure above the formation to prevent formation fluid entering the well. In such a conventional "overbalanced" well, the invasion of fluid is considered a kick, and if the well is not shut-in it can lead to a blowout, a dangerous situation. In underbalanced drilling, however, there is a "rotating head" at the surface - essentially a seal that diverts produced fluids to a separator while allowing the drill string to continue rotating.

If the formation pressure is relatively high, using a lower density mud will reduce the well bore pressure below the pore pressure of the formation. Sometimes an inert gas is injected into the drilling mud to reduce its equivalent density and hence its hydrostatic pressure throughout the well depth. This gas is commonly nitrogen, as it is non-combustible and readily available, but air, reduced oxygen air, processed flue gas and natural gas have all been used in this fashion.

Kinds of Underbalanced Drilling

There are several kinds of underbalanced drilling. The most common are listed below.

- Dry Air. This is also known as dusting. Here air compressors combined with a booster (which takes the head from the compressors and increases the pressure of the air, but does not increase the volume of air going down hole) are used and the only fluid injected into the well is a small amount of oil to reduce corrosion.

- Mist. A small amount of foaming agent (soap) is added into the flow of air. Fine particles of water and foam in an atmosphere of air bring cuttings back to the surface.
- Foam. A larger amount of foaming agent is added into the flow. Bubbles and slugs of bubbles in an atmosphere of mist bring cuttings back to the surface.
- Stable foam. An even larger amount of foaming agent is added into the flow. This is the consistency of a shaving cream.
- Airlift. Slugs and bubbles of air in a matrix of water, soap can or can not be added into the fluid flow of air.
- Aerated Mud. Air or another gas is injected into the flow of drilling mud. Degassing units are required to remove air before it can be recirculated.

Advantages

Underbalanced wells have several advantages over conventional drilling including:

- Eliminated formation damage. In a conventional well, drilling mud is forced into the formation in a process called invasion, which frequently causes formation damage - a decrease in the ability of the formation to transmit oil into the wellbore at a given pressure and flow rate. It may or may not be repairable. In underbalanced drilling, if the underbalanced state is maintained until the well becomes productive, invasion does not occur and formation damage can be completely avoided.
- Increased Rate of Penetration (ROP. With less pressure at the bottom of the wellbore, it is easier for the drill bit to cut and remove rock.
- Reduction of lost circulation. Lost circulation is when drilling mud flows into the formation uncontrollably. Large amounts of mud can be lost before a proper mud cake forms, or the loss can continue indefinitely. If the well is drilled underbalanced, mud will not enter the formation and the problem can be avoided.
- Differential sticking is eliminated. Differential sticking is when the drill pipe is pressed against the wellbore wall so that part

of its circumference will see only reservoir pressure, while the rest will continue to be pushed by wellbore pressure. As a result the pipe becomes stuck to the wall, and can require thousands of pounds of force to remove, which may prove impossible. Because the reservoir pressure is greater than the wellbore pressure in UBD, the pipe is pushed away from the walls, eliminating differential sticking.

- Formation damage Some rock formation have a reactive tendency to water. When drillmud is used the water in the drill mud reacts with the formation (mostly clay) and inheriently causes a formation damage (reduction in permeability and porosity) Use of underbalanced drilling can prevent it

Disadvantages

Underbalanced drilling is usually more expensive than conventional drilling (when drilling a deviated well which requires directional drilling tools), and has safety issues of its own. Technically the well is always in a blowout condition unless a heavier fluid is displaced into the well. Air drilling requires a faster up hole volume as the cuttings will fall faster down the annulus when the compressors are taken off the hole compared to having a higher viscosity fluid in the hole. Because air is compressible mud pulse telemetry measurement while drilling (MWD) tools which require an incompressible fluid can not work. Common technologies used to eliminate this problem are either electromagnetic MWD tools or wireline MWD tools. Downhole mechanics are usually more violent also because the volume of fluid going through a downhole motor or downhole hammer is greater than an equivalent fluid when drilling balanced or over balanced because of the need of higher up hole velocities. Corrosion is also a problem, but can be largely avoided using a coating oil or rust inhibitors.

Differential Sticking

Differential sticking is a problem that occurs when drilling a well with a greater well bore pressure than formation pressure, as is usually the case. The drill pipe is pressed against the wellbore wall so that part of its circumference will see only reservoir pressure, while the rest will continue to be pushed by wellbore pressure. As a result the pipe becomes stuck to the wall, and can require millions of pounds of force to remove, which may prove impossible. In many cases the drilling fluid (mud) weight is simply reduced, thus relieving the pressure difference and releasing the stuck pipe string. Should this option be unavailable, as in sour gas wells, a speciality fishing company is called to retrieve the stuck pipe or 'fish'. Many options exist once a fishing company is on site: oil or nitrogen may be pumped down the well, or the fish may be 'washed over' using a carbide shoe on a string of washpipe. Jarring is not usually attempted with differential sticking due to the massive amount of pressure that holds the pipe in place.

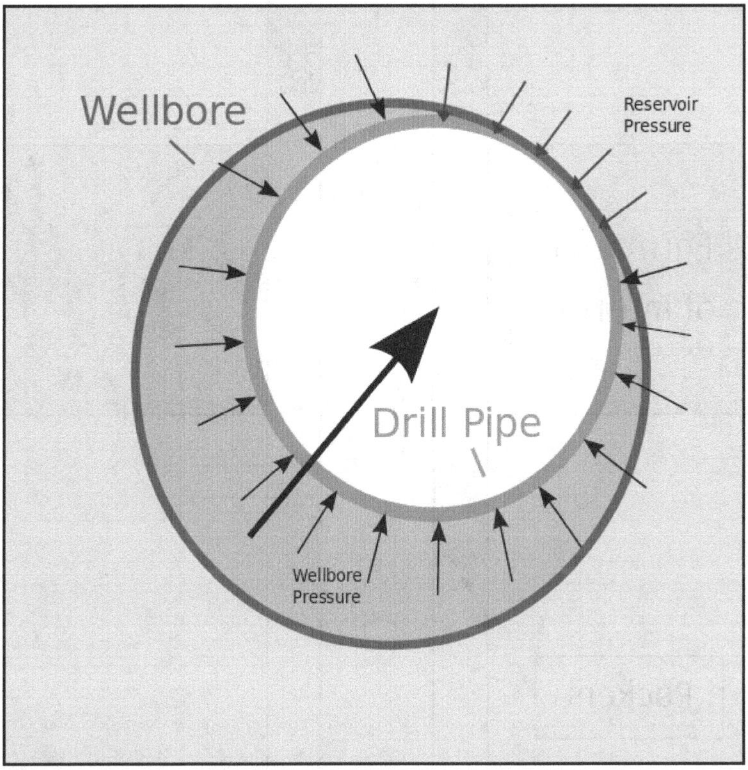

***Figure:** A diagram showing forces at work during differential sticking. The small black arrows represent pressure exerted on the drill pipe from the wellbore, the red arrows represent pressure exerted on the pipe from the formation (smaller than in the wellbore) and the large black arrow represents the net force on the pipe, which is pushing it into the wall.*

Drill Stem Test

A drill stem test (DST) is a procedure for isolating and testing the pressure, permeability and productive capacity of a geological formation during the drilling of a well. The test is an important measurement of pressure behaviour at the drill stem and is a valuable way of obtaining information on the formation fluid and establishing whether a well has found a commercial hydrocarbon reservoir.

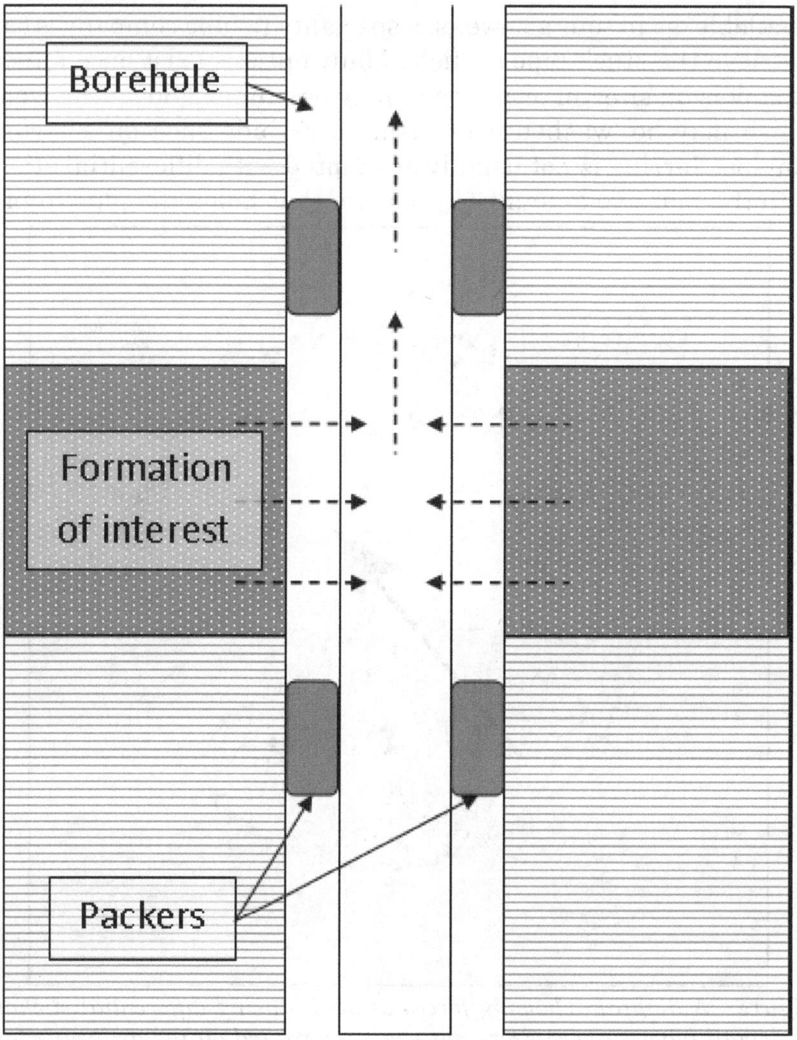

Figure: *Drill Stem Test*

History

Working in El Dorado, Arkansas, in the 1920s, E.C. Johnston and his brother M.O. Johnston developed the first drill stem tester and ran the first commercial drill stem test in 1926. In April 1929, the Johnston Formation Testing Corporation was granted a patent (U.S. Patent 1,709,940) and they subsequently refined the testing system in the early 1930s.

In the 1950s, Schlumberger introduced a method for testing formations using wireline. The Schlumberger formation-testing tool, placed in operation in 1953, fired a shaped charge through a rubber pad that had been expanded in the hole until it was securely fixed in the hole at the depth required. Formation fluids flowed through the perforation and connecting tubing into a container housed inside the tool. When filled, the container was closed, sealing the fluid sample at the formation pressure. The tool was then brought to the surface, where the sample could be examined. In 1956, Schlumberger acquired Johnston Testers and continues to perform drill stem tests and wireline formation tests in both open and cased holes.

Procedure

During normal well drilling, drilling mud is pumped through the drill stem and out of the drill bit. In a drill stem test, the drill bit is removed and replaced with the DST tool and devices are inflated above and below the section to be tested. These devices are known as packers and are used to make a seal between the borehole wall and the drill pipe, isolating the region of interest. A valve is opened, reducing the pressure in the drill stem to surface pressure, causing fluid to flow out of the packed-off formation and up to the surface.

Results

In a low permeability or low pressure formation, surface production may not be achieved but the volume and flow rate of fluid can still be analysed within the drill stem.

Lost Circulation

In oil or gas well drilling, lost circulation occurs when drilling fluid, known commonly as "mud", flows into one or more geological formations instead of returning up the annulus. Lost circulation can be a serious problem during the drilling of an oil well or gas well.

Consequences

The consequences of lost circulation can be as little as the loss of a few dollars of drilling fluid, or as disastrous as a blowout and loss of life, so close monitoring of tanks, pits, and flow from the well, to quickly assess and control lost circulation, is taught and practiced. If the amount of fluid in the wellbore drops due to lost circulation (or any other reason), hydrostatic pressure is reduced, which can allow a gas or fluid which is under a higher pressure than the reduced hydrostatic pressure to flow into the wellbore.

Another consequence of lost circulation is *dry drilling*. Dry drilling occurs when fluid is completely lost from the well bore without actual drilling coming to a stop. The effects of dry drilling range from as minor as destroying a bit to as serious as major damage to the wellbore requiring a new well to be drilled. Dry drilling can also cause severe damage to the drill string, including snapping the pipe, and the drilling rig itself.

Categories

Lost circulation falls into two main categories:

- "Minor losses" - losses are between 6 and 470 barrels (1 to 75 m^3), and remains within those amounts, or is ceased, within 48 hours.
- "Severe losses" - losses are greater than 470 barrels (75 m^3), or it takes greater than 48 hours to control or cease the lost circulation.

Total losses may also occur, where the return of fluids is completely lost to the surface. This may fall into either the minor or severe losses categories, depending on the amount lost and the time involved in regaining circulation.

Control

Although preferred, stopping lost circulation completely is not always possible or required. Controlled losses allow drilling to continue while keeping the wellbore full, preventing an influx of gas or fluid into the wellbore, known as a" kick", which can lead to a blowout.

A number of options are available when lost circulation occurs, depending on the severity.

- Minor losses may be controlled by increasing the viscosity of the fluid with bentonite and/or polymers, or with the addition of other additives, which typically includes sawdust.
- Severe losses will require increasing the viscosity of the fluid with bentonite and/or polymers and the addition of other additives, which typically includes sawdust.

Total losses can be regained through conventional use of increased viscosity and additives, or through use of unconventional methods such as pumping of golf balls, tree branches, rags, additive sacks, and other items in conjunction with, or followed by, a high viscosity fluid. If total losses occur and circulation cannot be regained, several options are available, depending on the operational requirements and depth being drilled in relation to desired production geological zones. Continuing drilling while pumping drilling fluid is one option, though continued drilling while pumping water is less costly and more often used. Sometimes the cuttings from continued drilling will aid in reducing leaks or stop losses altogether.

A third option is to cement the zone where the losses occurred, and to drill through the cement and continue drilling the well. This third option is very often the most cost effective if severe losses occur, as lost circulation sometimes cannot be controlled with other methods.

Additives

- The most common additive used to control or stop lost circulation is bentonite, which will seal small holes or fractures. Bentonite, in higher concentrations, increases viscosity, and therefore slows the fluid flow into the surrounding rock. Although bentonite is the most common additive used, it sometimes is not used at all, depending on the fluid being used and the depth of drilling in relation to desired production zones.
- Polymers are also sometimes used to increase the viscosity. Though these are more costly, they are more compatible with several types of fluid systems.
- There is a great variety of additives which physically plug or seal the losses, including sawdust, flaked cellophane, and crushed or ground gypsum.
- Other common and cheaper additives are shredded newspaper and cotton seed hulls. Cotton seed hulls are less preferred as they may cause wear to pump swabs and springs. Both of these are generally only used when either fresh or brine water is being used for the drilling fluid.
- When drilling in salt formations, brine water is typically used, as it does not as readily dissolve salt, preventing the formation of *washouts*. Washouts not only contribute to loss of circulation, but can jeopardize the integrity of the wellbore itself.

Additive Considerations

Several factors are considered in what additives are used:

- Hole size currently being drilled.
- Drilling fluid in use. The additives must be compatible.
- Depth of the well in regards to geological stability.
- Depth of well in relation to the desired production zones. Plugging a production zone is not a desired outcome.

- Drill bit nozzles sizes. If the additive(s) will not go through the drill bit, they cannot be used.
- Other drill string mechanical equipment such as a mud motor or MWD tools. If the additive(s) will not go through the drill string, they cannot be used.

Tracer Use in the Oil Industry

Tracers are used in the oil industry in order to qualitatively or quantitatively gauge how fluid flows through the reservoir, as well as being a useful tool for estimating residual oil saturation. Tracers can be used in either interwell tests or single well tests. In interwell tests, the tracer is injected at one well along with the carrier fluid (water in a waterflood or gas in a gasflood) and detected at a producing well after some period of time, which can be anything from days to years. In single well tests, tracer is injected into the formation from a well and then produced out the same well. The delay between a tracer that does not react with the formation (a conservative tracer) and one that does (a partitioning tracer) will give an indication of residual oil saturation, a piece of information that is difficult to acquire by other means.

Tracers can be radioactive or chemical, gas or liquid and have been used extensively in the oil industry and hydrology for decades.

Gamma Ray Logging

Gamma ray logging is a method of measuring naturally occurring gamma radiation to characterize the rock or sediment in a borehole or drill hole. It is a wireline logging method used in mining, mineral exploration, water-well drilling, for formation evaluation in oil and gas well drilling and for other related purposes. Different types of rock emit different amounts and different spectra of natural gamma radiation. In particular, shales usually emit more gamma rays than other sedimentary rocks, such as sandstone, gypsum, salt, coal, dolomite, or limestone because radioactive potassium is a common component in their clay content, and because the cation exchange capacity of clay causes them to adsorb uranium and thorium. This difference in radioactivity between shales and sandstones/carbonate rocks allows the gamma tool to distinguish between shales and non-shales.

The gamma ray log, like other types of well logging, is done by lowering an instrument down the drill hole and recording gamma radiation variation with depth. In the United States, the device most commonly records measurements at 1/2-foot intervals. Gamma radiation is usually recorded in API units, a measurement originated by the petroleum industry. Gamma logs are attenuated by diameter of the borehole because of the properties of the fluid filling the borehole, but because gamma logs are most often used in a qualitative way, corrections are usually not necessary.

Three elements and their decay chains are responsible for the radiation emitted by rock: potassium, thorium and uranium. Shales often contain potassium as part of their clay content, and tend to absorb uranium and thorium as well. A common gamma-ray log records the total radiation, and cannot distinguish between the radioactive elements, while a spectral gamma ray log can.

For standard GR logs the value measured is calculated from thorium in ppm, Uranium in ppm and potassium in percent. GR API = 8 × Uranium concentration in ppm + 4 × thorium concentration in ppm + 15 × potassium concentration in percent. Due to the weight of uranium concentration in the calculation anomalous concentrations of uranium can cause clean sand reservoirs to appear shaley. Spectral Gamma ray is used to provide an individual reading for each element so anomalies in concentration can be found and interpreted.

An advantage of the gamma log over some other types of well logs is that it works through the steel and cement walls of cased boreholes. Although concrete and steel absorb some of the gamma radiation, enough travels through the steel and cement to allow qualitative determinations.

Sometimes non-shales also have elevated levels of gamma radiation. Sandstone can contain uranium mineralization, potassium feldspar, clay filling, or rock fragments that cause it to have higher-than usual gamma readings. Coal and dolomite may contain absorbed uranium. Evaporite deposits may contain potassium minerals such as carnallite. When this is the case, spectral gamma ray logging can be done to identify these anomalies.

Spectral Logging

Some specialized gamma radiation logging distinguishes the three component decay chains (potassium, uranium, and thorium) by the wavelengths of their characteristic gamma emissions.

The characteristic gamma ray line that is associated with each component:

- Potassium : Gamma ray energy 1.46 MeV
- Thorium series: Gamma ray energy 2.62 MeV
- Uranium-Radium series: Gamma ray energy 1.76 MeV

Another example of the use of spectral gamma ray logs is to identify specific clay types, like Kaolinite or Illite. This can be used for environmental interpretation as Kaolinite forms from Feldspars in tropic soils by leaching of Potassium; and low Potassium readings may thus indicate paleosols. The identification of clay types is also useful for calculating the effective porosity of reservoir rock.

Use in Mineral Exploration

Gamma ray logs are also used in mineral exploration, especially exploration for phosphates, uranium, and potassium salts.

Resistivity Logging

Resistivity logging is a method of well logging that works by characterizing the rock or sediment in a borehole by measuring its electrical resistivity. Resistivity is a fundamental material property which represents how strongly a material opposes the flow of electric current. In these logs, resistivity is measured using 4 electrical probes to eliminate the resistance of the contact leads. The log must run in holes containing electrically conductive mud or water. Resistivity logging is sometimes used in mineral exploration and water-well drilling, but most commonly for formation evaluation in oil- and gas-well drilling. Most rock materials are essentially insulators, while their enclosed fluids are conductors. Hydrocarbon fluids are an exception, because they are almost infinitely resistive. When a formation is porous and contains salty water, the overall resistivity will be low. When the formation contains hydrocarbon, or contains very low porosity, its resistivity will be high. High resistivity values may indicate a hydrocarbon bearing formation.

Usually while drilling, drilling fluids invade the formation, changes in the resistivity are measured by the tool in the invaded zone. For this reason, several resistivity tools with different investigation lengths are used to measure the formation resistivity. If water based mud is used and oil is displaced, "deeper" resistivity logs (or those of the "virgin zone") will show lower conductivity than the invaded zone. If oil based mud is used and water is displaced, deeper logs will show higher conductivity than the invaded zone. This provides not only an indication of the fluids present, but also, at least quantitatively, whether the formation is permeable or not.

Use in Mineral Exploration

Resistivity logs are also used in mineral exploration, especially exploration for iron and potassium.

Density Logging

Density logging is a well logging tool that can provide a continuous record of a formation's bulk density along the length of a borehole. In geology, bulk density is a function of the density of the minerals forming a rock (i.e. matrix) and the fluid enclosed in the pore spaces. This is one of three well logging tools that are commonly used to calculate porosity, the other two being sonic logging and neutron porosity logging

History & Principle

The tool was initially developed in the 1950s and was in use throughout the hydrocarbon industry by the 1960s. A type of active nuclear tool, a radioactive source and detector are lowered down the borehole and the source emits medium-energy gamma rays into the formation. These gamma rays interact with electrons in the formation and are scattered in an interaction known as Compton scattering. The number of scattered gamma rays that reach the detector, placed at a set distance from the emitter, is related to the formation's electron density, which itself is related to the formation's bulk density (ρ_{bulk}) via

$$\rho_e = 2\rho_{bulk} \frac{Z}{A}$$

where Z is the atomic number, and A is the molecular weight of the compound. For most elements Z / A is about 1/2 (except for hydrogen where this ratio is 1). The electron density (ρ_e) in g/cm^3 determines the response of the density tool.

General Tool Design

The tool itself initially consisted of a radioactive source and a single detector, but this configuration is susceptible to the effects of the drilling fluid. In a similar way to how the sonic logging tool was improved to compensate for borehole effects, density logging now conventionally uses 2 or more detectors. In a 2 detector configuration, the short-spaced detector has a much shallower depth of investigation than the long-spaced detector so it is used to measure the effect that the drilling fluid has on the gamma ray detection. This result is then used to correct the long-spaced detector.

Inferring Porosity From Bulk Density

Assuming that the measured bulk density (ρ_{bulk}) only depends on matrix density (ρ_{matrix}) and fluid density (ρ_{fluid}), and that these values are known along the wellbore, porosity (ϕ) can be inferred by the formula

$$\phi = \frac{\rho_{matrix} - \rho_{bulk}}{\rho_{matrix} - \rho_{fluid}}$$

Common values of matrix density ρ_{matrix} (in g/cm^3) are:

- Quartz sand - 2.65
- Limestone - 2.71
- Dolomite - 2.87

This method is the most reliable porosity indicator for sandstones and limestones because their density is well-known. On the other hand, the density of clay minerals such as mudstone is highly variable, depending on depositional environment, overburden pressure, type of clay mineral and many other factors. It can vary from 2.1 (Montmorillonite) to 2.76 (Chlorite) so this tool is not as useful for determining their porosity. A fluid bulk density ρ_{fluid} of 1 g/cm^3 is appropriate where the water is fresh but highly saline water has a slightly higher density and lower values should be used for hydrocarbon reservoirs, depending on the hydrocarbon density and residual saturation.

In some applications hydrocarbons are indicated by the presence of abnormally high log porosities.

Sonic Logging

Sonic logging is a well logging tool that provides a formation's interval transit time, designated as Δt, which is a measure of a formation's capacity to transmit seismic waves. Geologically, this capacity varies with lithology and rock textures, most notably decreasing with an increasing effective porosity. This means that a sonic log can be used to calculate the porosity of a formation if the seismic velocity of the rock matrix, V_{mat}, and pore fluid, V_l, are known, which is very useful for hydrocarbon exploration.

Process of Sonic Logging

The velocity is calculated by measuring the travel time from the piezoelectric transmitter to the receiver. To compensate for the variations in the drilling mud thickness, there are actually two receivers, one near and one far. This is because the travel time within the drilling mud will be common for both, so the travel time within the formation is given by:

$$\Delta t = t_{far} - t_{near};$$

where t_{far} = travel time to far receiver; t_{near} = travel time to near receiver.

If it is necessary to compensate for tool tilt and variations in the borehole width then both up-down and down-up arrays can be used and an average can be calculated. Overall this gives a sonic log that can be made up of 1 or 2 pulse generators and 2 or 4 detectors, all located in single unit called a "sonde", which is lowered down the well. An additional way in which the sonic log tool can be altered is increasing or decreasing the separation between the source and receivers. This gives deeper penetration and overcomes the problem of low velocity zones posed by borehole wall damage.

Cycle Skipping

The returning signal is a wavetrain and not a sharp pulse, so the detectors are only activated at a certain signal threshold. Sometimes, both detectors won't be activated by the same peak (or trough) and the next peak (or trough) wave will activate one of them instead. This type of error is called cycle skipping and is easily identified because the time difference is equal to the time interval between successive pulse cycles.

Calculating Porosity

Many relationships between travel time and porosity have been proposed, the most commonly accepted is the Wyllie time-average equation. The equation basically holds that the total travel time recorded on the log is the sum of the time the sonic wave spends travelling the solid part of the rock, called the rock matrix and the time spent travelling through the fluids in the pores. This equation is empirical and makes no allowance for the structure of the rock matrix or the connectivity of the pore spaces so extra corrections can often be added to it. The Wyllie time-average equation is:

$$\frac{1}{V} = \frac{\phi}{V_f} + \frac{1-\phi}{V_{mat}}$$

where V= seismic velocity of the formation; V_f = seismic velocity of the pore fluid; V_{mat} = seismic velocity of the rock matrix; ϕ = porosity.

Accuracy

The accuracy of sonic logs is rather poor, evident by the fact that regular- and long-spaced log measurements often conflict, and this should be taken into account when there are disagreements between seismic data and sonic log data. Other considerations include that the resolution of sonic logs is on a scale of inches whereas seismic reflection data resolution is on the scale of metres, and the two methods use significantly different frequency ranges so travel times will vary due to dispersion. In order to investigate how the varying size of a borehole has affected a sonic log, the results can be plotted against those of a caliper log.

Use in Mineral Exploration

Sonic logs are also used in mineral exploration, especially exploration for iron and potassium.

Caliper Log

A caliper log is a well logging tool that provides a continuous measurement of the size and shape of a borehole along its depth and is commonly used in hydrocarbon exploration when drilling wells. The measurements that are recorded can be an important indicator of cave ins or shale swelling in the borehole, which can affect the results of other well logs.

Process

The caliper tool measures the variation in borehole diametre as it is withdrawn from the bottom of the hole, using two or more articulated arms that push against the borehole wall. Each arm is typically connected to a potentiometer which causes the resistance to change as the diameter of the borehole changes, creating a varying electrical signal that represents the changing shape of the borehole. This variation in output is translated into changes of diameter after a simple calibration and the caliper log is printed as a continuous series of values of hole diameter with depth.

Accuracy

Known challenges with caliper logging include borehole spiralling. The position of the drill bit may process as it drills, leading to spiralling shapes in the wellbore wall, as if the hole had been drilled by a screw. If the arms of the calliper log follow the grooves of the spiral, it will report too high an average diameter. Moving in and out of the grooves, the calliper will give erratic or periodically varying readings.

In most cases, the borehole's circumference will not be a perfect circle and therefore a calliper tool with several arms is required to obtain a true understanding of the size and shape of the borehole. The borehole can change to an oval shape after drilling, which can cause the calliper log to overestimate the size of the borehole if its arms get stuck axis.

Nuclear Magnetic Resonance Logging

Nuclear magnetic resonance (NMR) logging is a type of wireline logging that uses the NMR response of a formation to directly determine its porosity and permeability, providing a continuous record along the length of the borehole.

Background

NMR logging exploits the large magnetic moment of hydrogen, which is abundant in rocks in the form of water. The NMR signal amplitude is proportional to the quantity of hydrogen nuclei present in a formation and can be calibrated to give a value for porosity that is free from lithology effects. Uniquely, a petrophysicist can also analyse the rate of decay of the NMR signal amplitude to obtain information on the permeability of the formation - a crucial quantity in hydrocarbon exploration.

Relationship of NMR Signal to Pore Size

The most important mechanism affecting NMR relaxation is grain-surface relaxation. Molecules in fluids are in constant Brownian motion, diffusing about the pore space and bouncing off the grain surfaces. Upon interaction with the grain surface, hydrogen protons can transfer some nuclear spin energy to the grain (contributing to T1 relaxation) or irreversibley dephase (contributing to T2 relaxation). Therefore the speed of relaxation most significantly depends on how often the hydrogen nuclei collide with the grain surface and this is controlled by the surface-to-volume ratio of the pore in which the nuclei are located. Collisions are less frequent in larger pores, resulting in a slower decay of the NMR signal amplitude and allowing a petrophysicist to understand the distribution of pore sizes.

Chapter 6

Well Test

In the petroleum industry, a well test is the execution of a set of planned data acquisition activities to broaden the knowledge and understanding of hydrocarbons properties and characteristics of the underground reservoir where hydrocarbons are trapped. The test will also provide information about the state of the particular well used to collect data.

The overall objective is identifying the reservoir's capacity to produce hydrocarbons, such as oil, natural gas and condensate.

Data gathered during the test period includes volumetric flow rate and pressure observed in the selected well. Outcomes of a well test, for instance flow rate data and gas oil ratio data, may support the well allocation process for an ongoing production phase, while other data about the reservoir capabilities will support reservoir management.

Scope and Definitions

There are many flavours of well tests and various ways to categorize test types by its objectives, however two main categories only by objectives, these are productivity tests and descriptive tests. According to *The Lease Pumper's Handbook* of Oklahoma Commission on Marginally Producing Oil and Gas Wells, there are four basic well test types: potential tests, daily tests, productivity tests, and gas oil ratio tests, the latter three in the broader productivity test category.

Test objectives will change throughout the different phases of a reservoir or oil field, from the exploration phase of wildcat and appraisal wells, through the field development phase and finally through the production phase, which may also have variations from the initial period of production to improved recovery by the end of the field lifecycle time.

Exploration Phase

Professionals working with reservoir modelling may get information about the rock permeability from core samples. Other sources of information to the model are well log data and seismic data, but such data are complementary only, and for example, seismic data is insufficient to interpret whether a structural trap has been sealed. Information from well tests will supplement the amount of information with flow rate data, pressure data, and other, which is needed to build a rich reservoir model. The main objective in the exploration phase is to assess the size of a reservoir and state with a given certainty whether it has the properties for commercial exploitation and shall contribute to accounting for available reserves.

Well testing taking place before permanent well completion is referred to as drill stem testing.

Field Development Phase

The reservoir model is further developed to support the field development planning and to advise for the optimal location for extra production wells to be drilled. Descriptive well tests are designed and performed in the new wells.

Field Production Phase

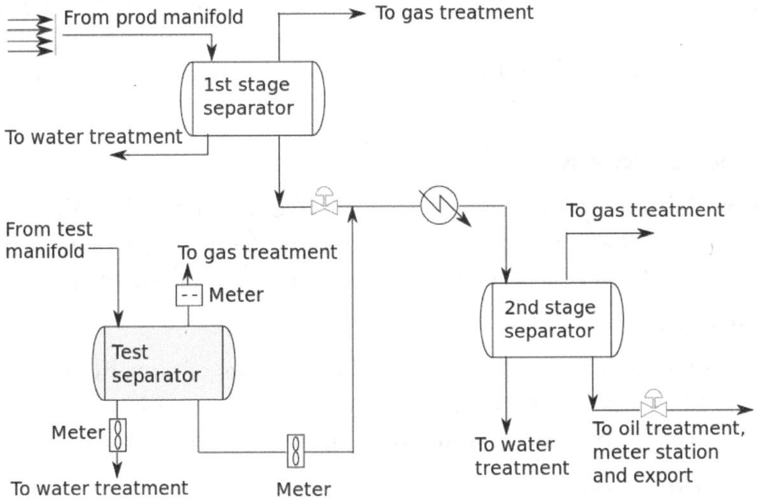

Figure: *Simplified process flow diagram. A plant is receiving a multi phased flow of oil and gas from many wells via a manifold. Flow from one well only may be taken to the test separator (shaded). The test separator has the feature to separate gas and water from the oil, and to have each component measured, under different conditions.*

Flow Test

This test has also been called daily test and may have various other namings. Often, and especially at offshore fields, a number of wells produce to a common separator, and flows from several separators or facilities may be headed into a commingled flow in pipeline that transports oil or gas for sale (export). The total flow rate of all wells in total are measured, but the contributions of the individual wells are unknown. It is important to know the individual contributions to account hydrocarbon material balance and for well monitoring and reservoir management.

To obtain individual well flow rates, it is common to use a smaller test separator. This is an isolated and down-scaled processing system in parallel with the normal flows. Regularly, for example once a month per well, the flow from one and only one selected well is led into the test separator for determining well flow rate for the selected well. The separator divides the flow from the well into the streams of individual products which typically are oil, gas and water, but may include natural-gas condensate. Contamination may also be removed and fluid samples collected.

This helps to allocate individual flow rate contributions, but the method has uncertainties. Flow rate, water cut, GOR and other parameters for the test system can deviate from production separators. This is generally taken into account by the allocation of products back to individual wells based on the field total, and by using data from the individual well tests.

Multiphase Flow Metre

A multiphase flow metre is a device used in the oil and gas industry to measure the individual phase flow rates of petroleum, water and gas mixtures produced during oil production processes.

Background

Knowledge of the individual fluid flow rates of a producing oil well is required to facilitate reservoir management, field development, operational control, flow assurance, and production allocation.

Conventional Solutions

Conventional solutions concerning two- and three-phase metering systems require expensive and cumbersome test separators, high maintenance, and field personnel intervention. These conventional solutions do not lend themselves to continuous monitoring or metering.

Moreover, with diminishing oil resources, oil companies are now frequently confronted with the need to recover hydrocarbons from marginally economical reservoirs. In order to ensure economic viability of these accumulations, the wells may have to be completed subsea, or crude oil from several wells sent to a common production facility with excess processing capacity. The economic constraints on such developments do not lend themselves to the continued deployment of three-phase separators as the primary measurement devices. Consequently, viable alternatives to three-phase separators are essential. Industry's response is the multiphase flow metre (MPFM).

Historical Development

The oil and gas industry began to be interested in developing MPFMs in the early 1980s. Prior to the 1980s, single-phase measurements alone were sufficient to meet the industry's needs. However, depleting oil reserves, along with smaller, deeper wells with higher water contents, saw the advent of increasingly frequent occurrences of multiphase flow where the single-phase metres were unable to cope. After a lengthy gestation period, MPFMs capable of performing the required measurements became commercially available. Since 1994, MPFM installation numbers have steadily increased as technology in the field has advanced, with substantial growth witnessed from 1999 onwards. A recent study estimated that there were approximately 2,700 MPFM applications including field allocation, production optimisation and mobile well testing in 2006.

A number of factors have instigated the recent rapid uptake of multiphase measurement technology: improved metre performances, decreases in metre costs, more compact metres enabling deployment of mobile systems, increases in oil prices and a wider assortment of operators. As the initial interest in multiphase flow metering came from the offshore industry, most of the multiphase metering activity was concentrated in the North Sea. However, the present distribution of multiphase flow metres is much more diverse.

Market

Industry experts have forecast that MPFMs will become feasible on an installation per well basis when their capital cost falls to around US$40,000 – US$60,000. The cost of MPFMs today remains in the range of US$100,000 – US$500,000 (varying with onshore/offshore, topside/subsea, the physical dimensions of the metre and the number of units ordered). Installation of these MPFMs can cost up to 25% of the hardware cost and associated operating costs are estimated at between US$20,000 and $40,000 per year. A number of novel multiphase metering techniques, employing a variety of technologies, have been developed which eliminate the need for three-phase separator deployment. These MPFMs offer substantial economic and operating advantages over their phase separating predecessor. Nevertheless, it is still widely recognised that no single MPFM on the market can meet all multiphase metering requirements.

Reservoir Simulation Tests the Limits of Built-in Post Processors

As computing power has increased, the volume of simulation data that reservoir engineers must work with has gone up exponentially. Working with so much important data, it's critical that engineers not miss anything. Vast amounts of money could be wasted if data is overlooked. And with all that data, it's very likely some could be.

For instance, before an oil company goes to the expense of producing a field, reservoir engineers are trusted to predict the optimum drilling location. These recommendations are based on increasing amounts of data When we developed Tecplot RS, we spoke with many reservoir engineers and learned a great deal about how they work, how they store their data, and how they import it with their modelling software.

XY Meets 3D

In watching engineers work, one issue we noticed was that they have to switch from XY to 3D views and can't see both together. Reservoir engineers have told us that being able to see both the XY and 3D view together makes a tremendous difference in the way they work and in their ability to understand the data.

With Tecplot RS, you can see your XY data and 3D data at the same time if you want. This ability gives engineers a new and faster way to understand the data they're seeing and make unexpected connections. Now, the software you use to understand reservoir data works the they way you want it to. Tecplot RS instantly lets you have a more complete understanding of your reservoir simulation data—all in a single environment. You can view:

%multiple runs simultaneously

%historical data

%geological models

The power of Tecplot RS is its ability to show you all your reservoir simulation data in a way that allows you to make more reliable predictions.

Managing Reservoirs of Reservoir Simulation Data

All the new computing power available to engineers is wonderful, but it's led to an exponential increase in simulation data to manage and try to understand. To handle and compare all that data, to sift through it, is a difficult task. And it's not just the recent data. Engineers looking at oil fields, for instance, are inputting data from as far back as 1950, when many of these oil fields began.

Tools like Tecplot RS make the management of this data easier with a more intuitive, simpler interface. You can open it up, start loading data and get meaningful results within about an hour.

Wellbore Storage

Wellbore Storage is not a factor on deep, dry gas, low perm well when testing from the surface! Wellbore storage can be a concern in wells with two characteristics: wells with a near wellbore limit, or wells with long storage times. For wells that have a near wellbore limit, wellbore storage may mask the pressure transient encountering this limit. This could lead to misinterpretation of the pressure transient test and erroneous characterization of the well. For wells with long storage times (greater than 24 hours for example), the length of the shut-in required to derive useful data from a build-up test may not be economically feasible. In situations like these, use of a downhole shut-in tool to reduce the duration of wellbore storage should be considered.

A wellbore storage question that we frequently receive concerns performing pressure transient tests from the surface on low perm wells. As outlined in a previous DRC Newsletter article, The Effect of Wellbore Storage on Surface Data, there are 3 primary factors that contribute to wellbore storage: Permeability of the producing formation, Compressibility of wellbore fluids and Volume.

There is nothing we can do to increase permeability or decrease the compressibility of produced fluids which leaves us with only VOLUME to manipulate. By running a downhole shut-in tool or when testing below a closed ball valve on a DST string you are dramatically decreasing wellbore volume and thus dramatically reducing the potential effects of wellbore storage on low permeability formations. It should be noted that downhole pressure gauges deployed on wireline (without a shut-in tool) and pressure taken from surface measurements would be equally affected by wellbore storage effects as the VOLUMES are the same.

Porosity Basics

Porosity is the volume of the non-solid portion of the rock filled with fluids, divided by the total volume of the rock.

Primary porosity is the porosity developed by the original sedimentation process by which the rock was created. In reports, it is often referred to in terms of percentages, while in calculations it is always a decimal fraction.

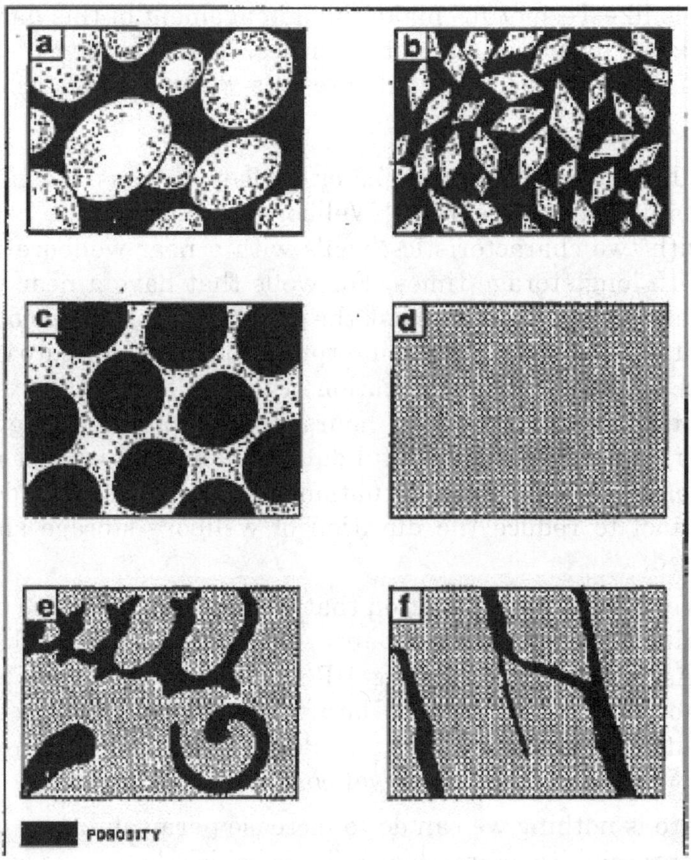

Figure: *Genetic classification of porosity type correlated with rock types: (a) intergranular (lime and dolomite grain stones, Rock Types 1 and 2; (b) Intercrystalline (sucrosic dolomites, Rock Type 3); (c) moldic (moldic colitic limestone and dolomite grainstones, Rock Type 4 and 5); (d) matrix or chalky (mudstones, chalks, Rock Type 6); (e) moldic or vuggy in addition to matrix (vuggy packstones and wackestones, Rock type 7); and (f) fracture or fissure porosity (Rock Type 8).*

Secondary porosity is created by processes other than primary cementation and compaction of the sediments. An example of secondary porosity can be found in the solution of limestone or dolomite by ground waters, a process which creates vugs or caverns. Fracturing also creates secondary porosity. Dolomitization results in the shrinking of solid rock volume as the material transforms from calcite to dolomite, giving a corresponding increase in porosity.

Porosity Types: a. intergranulas, b. Sucrosic,

c. Moldic or Oolitic, d. Matrix or Chalky,

e. Moldic or Vuggy, e. Fracture porosity.

Intergranular porosity is primary, the

balance are secondary.

Log analysts define porosity somewhat differently, due to the nature of the measuring techniques.

Porosity in Sandstones

We tend to think of sandstones as being composed of quartz grains, but this is a false impression based on too many idealized cinematic beaches and cast-away island songs. Most sandstones are made of many different minerals; some have no quartz at all. So in the following discussion, please think of sand grains as being composed of a variety of minerals, not necessarily pure quartz.

To acquire an appreciation for the values of porosity generally encountered, assume round balls of the same size are stacked on top of each other in columns. Calculations will show a porosity of 47.6%. Spherical sand grains 1/10 the size of the balls stacked one on top of the other will have the same porosity, 47.6%.

If the same balls are packed in the closest possible arrangement in which the upper ball sits in the valley between the four lower balls, the porosity is reduced to 25.9%. Again, changing the size of the balls will not change the porosity as long as all the balls are the same size.

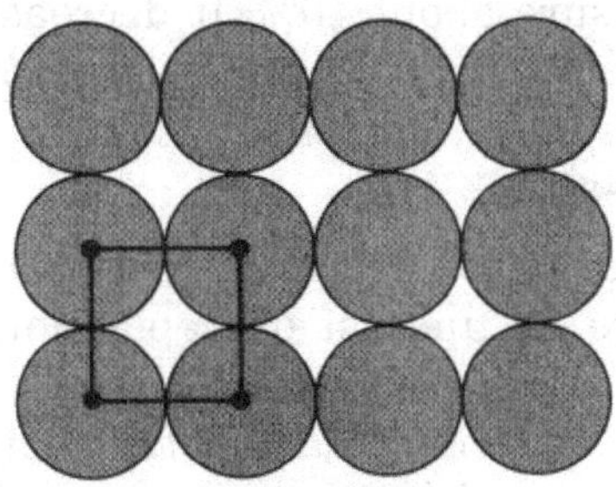

Cubic packing
(48% porosity)

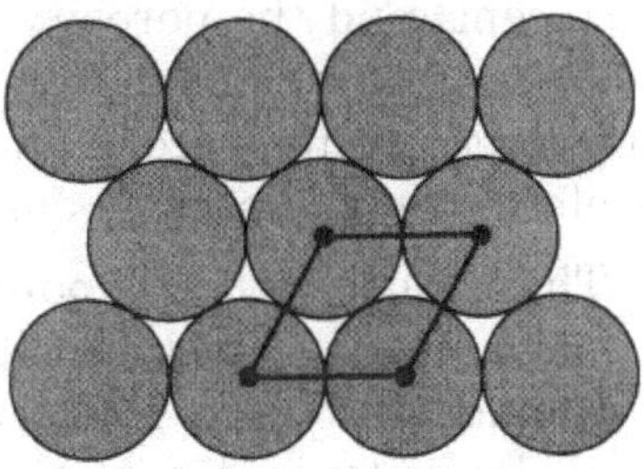

Rhombohedral packing
(26% porosity)

Cubic packing 47% porosity (left) Rhombic packing 26% porosity (right)

Mixing the sizes of the balls will create lower porosity, since small ones can fit in spaces created between the larger ones. The term "sorting" is used to describe the distribution of grain sizes in a sandstone. Very well sorted rocks have fairly uniform grain size and high porosity. Poorly sorted sands have a wide range of grain size and poor porosity, illustrated below. Grain size classifications are shown on the scale at right, below.

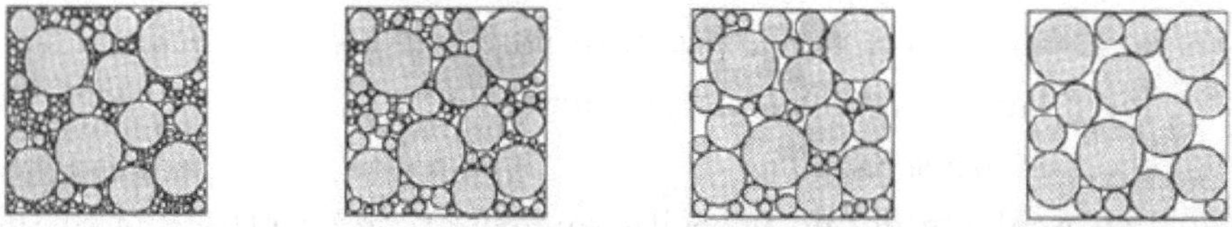

Poorly Sorted Low Porosity	Moderately Sorted Poor Porosity	Well Sorted Good Porosity	Very Well Sorted Excellent Porosity

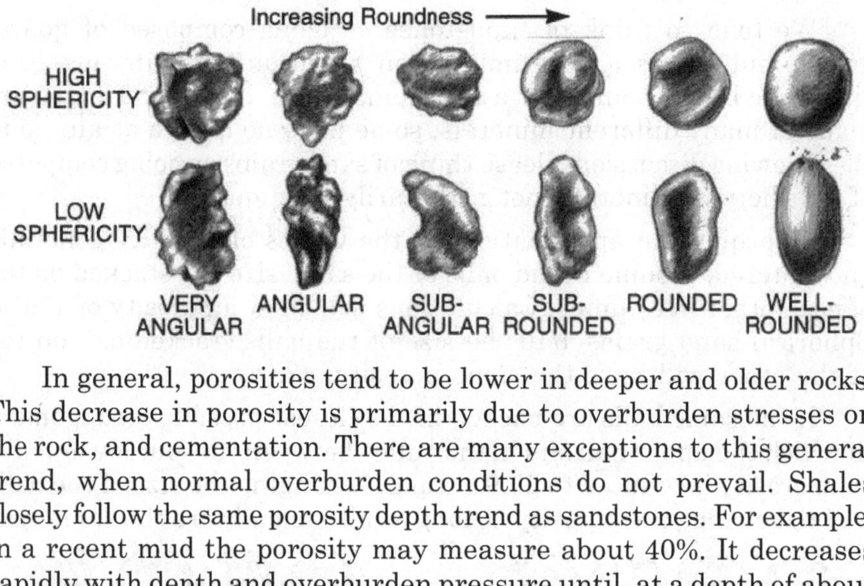

In general, porosities tend to be lower in deeper and older rocks. This decrease in porosity is primarily due to overburden stresses on the rock, and cementation. There are many exceptions to this general trend, when normal overburden conditions do not prevail. Shales closely follow the same porosity depth trend as sandstones. For example, in a recent mud the porosity may measure about 40%. It decreases rapidly with depth and overburden pressure until, at a depth of about 10,000 feet, normal porosities are less than 5%. Shales are plastic and therefore, compress more easily than sands.

These basic trends of porosity versus depth are not as noticeable in carbonates, where porosity is more a function of depositional environment and secondary processes, both unrelated to depth of burial. Porosity in a real shale is not effective; that is, the water cannot move as quickly as in a sandstone with the same apparent porosity. Water in shale can be expelled over large geologic time periods, but it will not flow in the usual sense of the word.

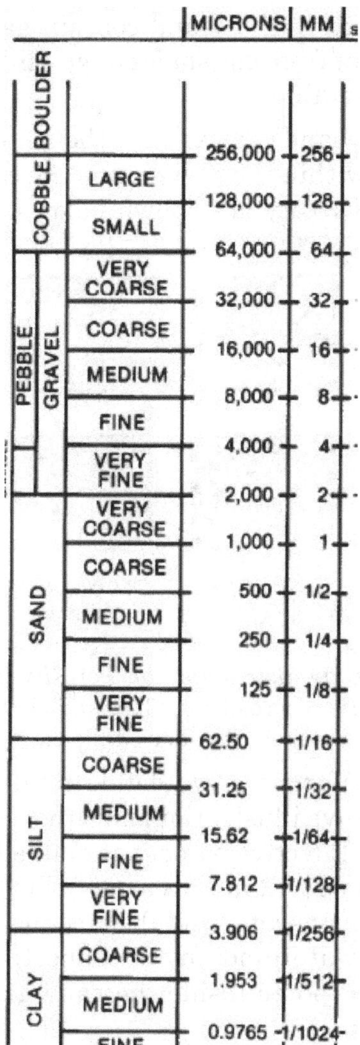

The highest porosity normally anticipated is 47.6%. A more probable porosity is in the mid-twenties. The normal range of porosities in granular systems is 5% to 35%.

However, many intervals that have been traditionally thought of as "shale" are really silty shales or sandy shales. These may have sufficient porosity to store hydrocarbons that might flow. This is especially true for gas, and many "gas shales" are silty shales with effective porosity. Other gas shales are mostly shale and gas is stored on the surface of kerogen within the shale. This is adsorbed gas. Laminated shaly sands are also called gas shales in some literature.

While they are definitely shaly and contain gas, the petrophysical model is quite different from gas shales or gas silts, so the "laminated" adjective should be retained.

Porosity in shaly sands varies with the amount and distribution of the clay minerals within the sandstone. The common distributions and their effect on porosity are shown below.

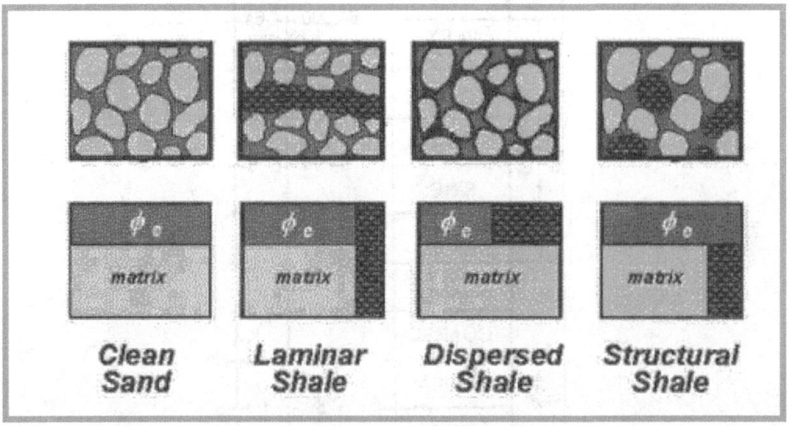

Figure: *The Effect of Clay distribution on Porosity in a Shaly Sand. Sand grains are yellow, effective* porosity is blue, and clay (including clay bound water) is coloured black.

For log analysis purposes, we define total porosity as the pore space (blue area in above illustration) plus the clay bound water (part of the black shading). Effective porosity is defined as the total porosity minus the clay bound water (blue area only). Further adjustments are sometimes made to generate useful or connected porosity, which excludes clay bound water and any unconnected pores, such as pin-point vugs or isolated pores inside the sand grains.

Useful Porosity

There is a recent trend among petrophysicists and engineers to partition porosity into a useful and a non-useful fraction. The concept of useful porosity, as opposed to effective porosity, is helpful where very small pores exist.

These tiny pores do not connect to other pores and thus do not contribute to useful reservoir volume or reservoir energy. They are invariably water filled and nothing flows from them or through them. The tiny pores are called micro porosity; the larger, more effective, pores are called macro porosity. Personally, I prefer the term connected and unconnected (or poorly connected) porosity, as illustrated below:

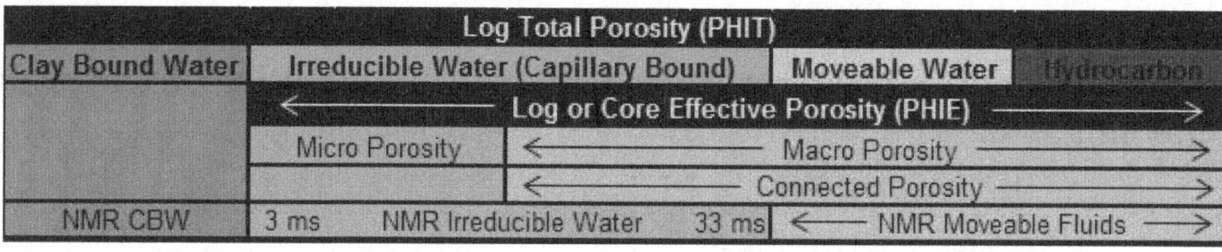

Figure: *Porosity definitions related to useful or connected porosity. Some micro porosity may not be observed in conventional core analysis. Most porosity indicating logs see unconnected porosity, but the sonic log may not see any or all of the microporosity.*

Thus:

1: PHIuse = PHIe - PHImicro

In sandstones, micro porosity is often associated with intraparticle porosity in volcanic rock fragments and kaolinite that are part of the sandstone mineral mixture. In carbonates, micro porosity is associated with micrite, matrix, fossil skeletons, or pin point vugs. Larger vugs are often connected.

The quantity of micro porosity cannot always be found directly from logs but is usually assessed as a constant fraction, KM1, of the effective porosity. This constant can be found by examination of thin section visual porosity. Where micro porosity is associated with silt or a volcanic mineral (Vmin2) in a quartz sandstone:

2: KM1 = Vsilt / (Vqrtz + Vsilt)

OR 2A: KM1 = Vmin2 / (Vqrtz + Vmin2)

3: PHIuse = PHIe * KM1

In some cases, the micro porosity is assumed to be a constant value instead of a constant fraction of the silt volume, PHIoffset, over an interval (ie, PHImicro is not proportional to effective porosity). This appears to happen in carbonates with unconnected pinpoint vugs (PHIppv), micritic carbonates (PHImict), or carbonates with matrix porosity (PHImatr). In all three cases, PHIoffset is found by comparing visual porosity in thin sections to log analysis porosity.

4: PHIuse = PHIe - PHIoffset

In log analysis terminology, matrix porosity usually means effective porosity (PHIe). However, in petrographic (thin section) analysis, matrix porosity (PHImatr) is non-useful porosity contained in the very fine-grained matrix material deposited between the granular or crystalline rock structure.

PHIppv, PHImict, and PHImatr may be varied according to rules developed by the analyst for the zone. A crossplot of visual porosity from thin section analysis versus PHIe from logs is a useful tool for determining the appropriate correction to obtain PHIuse. Typical rules might be:

5: PHIuse = PHIe – PHIsec

6: PHIuse = PHIsec

7: PHIuse = PHIe – KMATR * (1 – PHIe) / (1 – KMATR)

8: PHIuse = PHIe – PHIsc * KMICT / PHISavg

KMATR and KMICT would be in the range 0.01 to 0.08, averaging 0.04, and cannot exceed PHIt.

Definitions of Porosity For Log Analysis Purposes

The above discussion covers the geological definitions of porosity. Petrophysicists, log analysts, and engineers use more specific terms based on the log analysis model described.

Here are the definitions that derive from that model.

***DFN* 1:** The formation rock/fluid model is comprised of:

- the matrix rock (Vrock)
- the pore space (or porosity) within the matrix rock (PHIe)
- the shale content of the matrix rock (Vsh)

By definition, Vrock + PHIe + Vsh = 1.00

***DFN* 2:** The matrix rock component (Veock) can be subdivided into two or more constituents

(Vmin1, Vmin2, ...), such as:

- limestone, dolomite, and anhydrite or
- quartz, calcite cement, and glauconite

The mineral mixture can be quite complex and log analysis may not resolve all constituents.

***DFN* 3:** The shale component (Vsh) can be classified further into:

- one or more clays (Vcl1, Vcl2, ...)
- silt (Vsilt)
- water trapped into the shale matrix due to lack of sufficient permeability to allow the water to escape
- water locked onto the surface of the clay minerals
- water absorbed chemically into the molecules of the clay minerals

The sum of the three water volumes is called clay bound water (CBW). CBW varies with shale volume and is zero when Vsh = 0.

By definition, Vsh = Vcl + Vsilt + CBW

***DFN* 4:** Bulk volume water of shale (BVWSH) is the sum of the three water volumes listed above in the definition of shale and is determined in a zone that is considered to be

100% shale.

By Definition, CBW = BVWSH * Vsh

***DFN* 5:** Total porosity (PHIt) is the sum of:

- clay bound water (CBW)
- free water, including irreducible water (BVW)
- hydrocarbon (BVH)

Some of the "free water" is not free to move - it is, however, not "bound" to the shale. It could also be called pore water.

***DFN* 6:** Effective porosity (PHIe) is the sum of:

- free water, including irreducible water (BVW)
- hydrocarbon (BVH)

***DFN* 7:** Effective porosity is the porosity of the reservoir rock, excluding clay bound water

(CBW).

PHIe = PHIt – CBW

OR PHIe = PHIt – Vsh * BVWSH

***DFN* 8:** Free water (BVW) is further subdivided into:

- a mobile portion free to flow out of the reservoir (BVWm)
- an immobile or irreducible water volume bound to the matrix rock by surface tension (BVI or BVWir)

BVI is sometimes called "bound water" or "capillary bound water", but this is confusing, so "irreducible water" is a better term.

***DFN* 9:** Hydrocarbon volume (BVH) can be classified into:

- mobile hydrocarbon (BVHm)
- residual hydrocarbon (BVHr)

***DFN* 10:** Free fluid index (FFI) is the sum of BVWm, BVHm, and BVHr. It is also called moveable

fluid (BVM) or useful porosity (PHIuse).

PHIuse = BVM = FFI = BVWm + BVHm + BVHr

OR PHIuse = PHIe – BVI

OR PHIuse = PHIe * (1 – SWir)

This definition is needed for older nuclear magnetic logs since they could not see BVWir.

Non-useful porosity occurs as tiny pores that do not connect to any other pores. They are almost invariably filled with immoveable water and do not contribute to useful reservoir volume or energy. Such pores occur in silt, volcanic rock fragments in sandstones, and in micritic, vuggy, or skeletal carbonates. The NMR may see some of this non-useful porosity; the jury is still out.

Porosity Indicating Logs

Porosity Indicating Logs

Sonic (DELT)
Density (DENS or PHID)
Neutron (PHIN)
Resistivity (RESS)
Electromagnetic Propagation (EPT)
Microlog (MLC)
Nuclear Magnetic (NMR)

There are numerous porosity indicating logs, as shown in the box at right, and many flavours of each, depending on the age, design, and logging environment. Generic analysis equations, based on the Log Response Equation, for each basic tool type. They will work for almost all available tool types. There may be rare occasions when a customized analysis model might be required.

All the porosity models require some assumptions about such things as fluid type and matrix rock properties. With the exception of the resistivity log formula, used for analysis of ancient logs, the methods involve corrections for the effects of shale.Various methods are presented in other sections of this Handbook to calculate porosity from individual or combinations of two or more logs. Two-log combinations are termed crossplot methods, since the log data can be plotted on the X and Y axes of a graph. Three or more log combinations require solution by simultaneous equations, and are usually done on a computer.

Shale corrections are applied to porosity logs to determine effective porosity. Since shale contains some water, this water must be subtracted from the total porosity as measured by conventional logging tools. The mathematical method for finding shale volume is the same for all the shale distribution types, but the method for applying the shale correction to the porosity varies.

Correcting for shale is only half the battle. The other half is to correct for the mineral composition of the rocks. In most carbonate reservoirs, the lithology is usually reasonably well known from sample descriptions or can be determined from log response, so this step is relatively straightforward.

This is not true in sandstones because the mineral makeup of the sand is NOT usually described in much detail. There is a universal trend to give sandstones the physical properties of pure quartz, but this is almost universally NOT appropriate. Most sandstones contain other minerals besides quartz, such as mica, volcanic rock fragments, calcite, dolomite, anhydrite, and ferrous minerals, as well as the shale and clay described above. All of these minerals have different log responses than quartz. If a sandstone is assumed to be pure quartz when it is not, the commonly used properties of quartz will provide a pessimistic porosity answer.

Thus, authors and service company manuals that present quartz properties for "sandstone" are misleading their audience into believing these properties are constant. In more than 40 years of petrophysical analysis, I have never seen a thin section or XRD report that gave an assay of 100% quartz in any petroleum reservoir. A 100% quartz sand is very rare. If anyone doubts this statement, look at the PEF curve. If it reads more than 1.8, you have "quartz plus other things" in your sandstone.

There is a story (it may even be true) that reserves for the early North Sea discoveries were seriously underestimated because the mica in the sands was not accounted for properly. The engineers used density log porosity without correcting for the real matrix density. If true, good engineering practice would have undersized all the offshore equipment and early cash flow and rate of return on investment would have been significantly reduced. If the myth that sandstone is pure quartz is perpetuated, there will be more economic blunders of this type.

To further confuse the uninitiated, many logs show data on a "porosity" scale. These log curves are transforms of some measured physical property to an approximate porosity based on some arbitrary parameters. Examples are density, neutron, or sonic porosity on so-called Sandstone, Limestone, or Dolomite porosity scales. Porosity as defined by these transforms is only directly useful if there is no shale, the scale matches the rock mineralogy and there are no accessory minerals. Real reservoirs are rarely this simple. DO NOT use these porosity transforms without further analysis unless all the arbitrary assumptions used to create them match exactly the rock you are analyzing. Some people call these porosity curves an "interpretation". They are not. They are merely a transform of the raw data to a more attractive scale. The difference between a transform and an interpretation is critical. Interpretation infers some intelligent thought went into creating and understanding the result. The service company running the log does not provide interpretations. YOU are the interpreter,

There are endless cases where a transform to an inappropriate porosity scale has caused millions in losses due to poorly informed analysts who see "gas cross over" when there is no gas, or who read porosity directly from the transform and either seriously over estimate or under estimate reservoir effective porosity.

Porosity from the Sonic Log

"Slowness" is the new word for sonic or acoustic travel time. The inverse of slowness is speediness or velocity. We continue to use travel time in this Handbook - it's hard to teach old dogs new tricks.

The basic relationship for the sonic log starts with the elastic constants of rocks:

Vp = KS4 * (Kc / DENS) ^ 0.5
Vs = KS4 * (N / DENS) ^ 0.5

Where:

Kc = composite bulk modulus of the rock

N = composite shear modulus of the rock

DENS = density of the rock

Vp = compressional velocity of the rock

Vs = shear velocity of the rock

KS4 = units conversion term

Since density and the moduli vary with the rock mineral type, shale volume, porosity, and the type of fluids in the pores, so does velocity. Travel time is the inverse of velocity, so it varies with these components as well. M. R. J. Wyllie plotted core porosity against travel time and found a linear relationship that varied in slope depending on mineralogy. Later a shale term was added. At low to medium porosity, a plot of velocity versus core porosity also appears to be linear, but the square root term in each of the above equations suggests that either arrangement should be a curve. Wyllie's linear approximation is sufficient for many situations and is very widely used.

The response equation for the sonic log follows the classical form:

1: DTC = PHIe * Sxo * DTCw (water term)
+ PHIe * (1 - Sxo) * DTCh (hydrocarbon term)
+ Vsh * DTCsh (shale term)
+ (1 - Vsh - PHIe) * Sum (Vi * DTCi) (matrix term)

WHERE

DTCh = log reading in 100% hydrocarbon
DTCi = log reading in 100% of the ith component of matrix rock
DTC = log reading
DTCsh = log reading in 100% shale
DTCw = log reading in 100% water
PHIe = effective porosity (fractional)
Sxo = water saturation in invaded zone (fractional)
Vi = volume of ith component of matrix rock
Vsh = volume of shale (fractional)

To solve for porosity from the sonic log, we assume DTCh, DTCi, DTCsh, DTCw, and Vsh are known. We also assume DTCw = DTCh and Sxo = 1.0 when no gas is present. If gas is indicated, we make assumptions about DELTh and Sxo, usually in the form of a correction factor to the gas free case. The usual result is:

The response equation is not rigorous and many exceptions are noted below. Mineral and fluid parameters are shown HERE. Shale properties are selected from the log in an obvious shale zone.

PHIsc - Porosity From the Sonic Log (Wyllie Method)

Calculate sonic compaction correction

2: KCP = max (1, DTCSH / KS9)

WHERE: KS9 = 100 for English units, 328 for Metric units

Calculate total sonic porosity

3: PHIS = (DTC – DTCMA) / (DTCW – DTCMA) / KCP

Correct sonic porosity for shale

4: PHISSH = (DTCSH - DTCMA) / (DTCW – DCTMA) / KCP

5: PHIsc = PHIS – Vsh * PHISSH

Correct sonic porosity for gas effect

6: IF SONICGASSWITCH$ = "ON"

7: THEN PHIsc = KS * PHIS

WHERE:

KCP = compaction factor (fractional)

DTC = sonic log reading in zone of interest (usec/ft or usec/m)

DTCMA = sonic log reading in 100% matrix rock (usec/ft or usec/m)

DTCSH = sonic log reading in 100% shale (usec/ft or usec/m)

DTCW = sonic log reading in 100% water (usec/ft or usec/m)

KS = sonic log gas correction factor

PHIS = porosity from sonic log (corrected for compaction if needed) (fractional)

PHIsc = porosity from sonic log by Wyllie method (fractional)

PHISSH = apparent sonic porosity of 100% shale after compaction correction (if needed) (fractional)

Vsh = shale volume (fractional)

Comments: Of the three "one-log" porosity methods, the sonic corrected for shale is the preferred one for wells drilled after 1957 and before 1965. However, crossplot methods or the density log corrected for shale are usually better if the log data is available.

Use the compaction correction only if CDTSH > 100 (for English units) or CDTSH > 328 (for Metric units). In western North America, this is normally required when above 3,000 - 4,000 feet (900 - 1,200m).

KS is in the range 0.7 to 1.0 depending on gas density invasion and local experience. It can be derived by comparing the calculated porosity with the true porosity from cores or density neutron crossplot methods.

Use gas correction only if PHIS is too high compared to other sources, only if the zone is clean and does not need shale corrections, and if gas is known to be present. The need for this correction is rare. It is very unlikely that a gas correction will be needed in shaly sands since invasion should be relatively deep.

Another way of making gas corrections is to change DELTW to a higher value, representing the travel time of sound in a mixture of gas and water. This value depends on water saturation in the invaded zone, pressure, temperature and gas compressibility. Values in the range of 600 usec/ft (1900 usec/m) at shallow depths to 300 usec/ft (950 usec/m) at 6000 feet (2000 metres) are recommended as a starting point.

KCP can be calculated if true porosity of a clean zone is known from core, neutron, or density log data:

8: KCP = PHIS / PHItrue

WHERE:

KCP = compaction factor (fractional) (usec/ft or usec/m)

PHIS = sonic log porosity in clean sand (fractional)

PHItrue = actual porosity in clean sand from core or density data (fractional)

NUMERICAL EXAMPLE:

1. Wyllie Method – Shaly Sand

DTC = 300 usec/m

DTCSH = 328 usec/m

DTCMA = 182 usec/m

DTCW = 616 usec/m

Vsh = 0.33

KCP = 1.0

Therefore compaction correction is not needed.

PHIS = (300 – 182) / (616 – 182) / 1.0 = 0.27

PHISSH = (328 – 182) / (616 – 182) / 1.0 = 0.34

PHIsc = 0.27 – 0.33 * 0.34 = 0.16

PHIsc is not too high, and no gas is known to be present. Hence, no gas correction is made.

2. Wyllie Method – Clean Gas Sand

DTC = 380 usec/m

DTCSH = 328 usec/m

DTCMA = 182 usec/m

DTCW = 616 usec/m

Vsh = 0.0

KCP = 1.0

PHIS = (380 – 182) / (616 – 182) / 1.0 = 0.46

PHISSH = (328 – 182) / (616 – 182) = 0.36

PHIsc = 0.46 – 0.0 * 0.36 = 0.46

PHIsc is too high due to gas effect – assume KS = 0.75

PHIsc = 0.75 * 0.46 = 0.33

3. Wyllie Method – Un-compacted Sand

DTC = 375 usec/m

DTCSH = 460 usec/m

DTCMA = 182 usec/m

DTCW = 616 usec/m

Vsh = 0.0

KCP = 460 / 328 = 1.40

PHIsc = PHIS = (375 – 182) / (616 – 182) / 1.40 = 0.31

No gas correction is required.

No shale correction is required.

Phishr - Porosity From the Sonic Log (Hunt-Raymer Method)

The Hunt-Raymer method is a newer formula which is a non-linear calibration of observed porosity versus log response data. It should be used in clean sands and carbonates only, or log data may be corrected for shale first. It can be used in un-compacted sands without the compaction correction described in the Wyllie method given above. The algorithm is derived from the following empirical relationship:

11: VELOG = VELMA * ((1 – PHIe) ^ 2) + VELW * PHIe

This can be solved for porosity in the following way:

Calculate sonic log reading corrected for shale:

12: DTC1 = DTC – Vsh * (DTCSH – DTCMA)

Calculate sonic porosity

13: C = DTCMA / (2 * DTCW)

14: PHIShr = 1 – C – (C ^ 2 – DTCMA / DTCW + DTCMA / DTCc) ^ 0.5

WHERE:

C = intermediate term

DTC = sonic log reading in zone of interest (usec/ft or usec/m)

DTC1 = sonic log reading corrected for shale (usec/ft or usec/m)

DTCMA = sonic log reading in 100% matrix rock (usec/ft or usec/m)

DTCSH = sonic log reading in 100% shale (usec/ft or usec/m)

DTCW = sonic log reading in 100% water (usec/ft or usec/m)

PHIShr = porosity from sonic log by Hunt-Raymer method (fractional)

VELOG = sonic velocity log reading (ft/sec or m/sec)

VELMA = sonic velocity log reading in 100% matrix (ft/sec or m/sec)

VELW = sonic velocity log reading in 100% water (ft/sec or m/sec)

Vsh = shale volume (fractional)

Although the original paper does not discuss shale corrections, they are essential. Gas corrections similar to those used in the Wyllie method can be used if needed. The answer porosity will be too high in gas if the corrections are not made. The method is not universally applicable and should be tested in each area before use.

Another way of making gas corrections is to change DELTW to a higher value, representing the travel time of sound in a mixture of gas and water. This value depends on water saturation in the invaded zone, pressure, temperature and gas compressibility. Values in the range of 600 usec/ft (1900 usec/m) at shallow depths to 300 usec/ft (950 usec/m) at 6000 feet (2000 metres) are recommended as a starting point.

Numerical Example

1. Hunt-Raymer Method - Shaly Sand

DTC = 300 usec/m

DTCSH = 328 usec/m

DTCMA = 182 usec/m

DTCW = 616 usec/m

Vsh = 0.33

KCP = 1.0

DTC1 = 300 – 0.33 * (328 – 182) = 251 usec/m

C = 182 / (2 * 616) = 0.147

PHIShr = 1 – 0.147 – (0.147 ^ 2 – 182 / 616 + 182 / 251) ^ 0.5 = 0.18

2. Hunt-Raymer Method – Clean Gas Sand

DTC1 = 380 – 0.00 * (328 – 182) = 380 usec/m

C = 182 / (2 * 616) = 0.147

PHIShr = 1 – 0.147 – (0.147 ^ 2 – 182 / 616 + 182 / 380) ^ 0.5 = 0.40

Porosity is too high due to gas effect – assume KS = 0.80.

PHIsc = 0.80 * 0.40 = 0.32

3. Hunt-Raymer Method – Un-compacted Sand

DTC1 = 375 – 0.33 * (460 – 182) = 375 usec/m

C = 182 / (2 * 616) = 0.147

PHIShr = 1 – 0.147 – (0.147 ^ 2 – 182 / 616 + 182 / 375) ^ 0.5 = 0.39

This result is a little high compared to the more conventional method.

PHIshear - POROSITY FROM THE DIPOLE SHEAR SONIC LOG (WYLLIE METHOD)

The newer sonic logs record shear travel time as well as the compressional travel tine. The compressional data is processed as discussed above under the Wyllie and Raymer-Hunt methods. Shear travel time can be used in the Wyllie equation, using fictitious values for fluid travel time.

There is very little fluid effect on shear data so there is no gas correction.

Calculate total sonic porosity

15: PHIS_S = (DTS – DTSMA) / (DTSW – DTSMA)

Correct sonic porosity for shale

16: PHISSH_S = (DTSSH – DTSMA) / (DTSW – DTSMA)

173: PHIsc_S = PHIS_S – Vsh * PHISSH_S

Where

DTS = shear sonic log reading in zone of interest (usec/ft or usec/m)

DTSMA = shear sonic log reading in 100% matrix rock (usec/ft or usec/m)

DTSSH = shear sonic log reading in 100% shale (usec/ft or usec/m)

DTSW = (fictitious) shear sonic log reading in 100% water (usec/ft or usec/m)

PHIS_S = porosity from shear sonic log before shale correction (fractional)

PHIsc_S = porosity from shear sonic log by Wyllie method (fractional)

PHISSH_S = apparent shear sonic porosity of 100% shale (fractional)

Vsh = shale volume (fractional)

Comments: Shear travel time is more sensitive to porosity than compressional data.

No gas correction is needed.

The measurement can usually be made through casing so this is a good choice for cased hole logging.

There is no record of a compaction correction being applied, but this may be needed. Comparison to core porosity or density neutron crossplot porosity will indicate when such a correction is needed.

Table: *RECOMMENDED PARAMETRES - SHEAR TRAVEL TIME*

	English - usec/ft	***Metric - usec/m***
DTSSH	96 - 240	490 - 770
DTSW fresh water	350	1280
salt water	340	1200
DTSMA		
Granite	80.0	262
Quartz	88.8	291
Limey sandstone	88.9	292
Limestone	89.9	294
Limey dolomite	82.3	270
Dolomite	74.8	245
Anhydrite	85.0	280
Coal	152+	500+

POROSITY FROM THE DENSITY LOG

The response equation for the density log in porosity units follows the classical form:

1: PHID = PHIe * Sxo * PHIDw (water term)
+ PHIe * (1 – Sxo) * PHIDh (hydrocarbon term)
+ Vsh * PHIDsh (shale term)
+ (1 – Vsh – PHIe) * Sum (Vi * PHIDi) (matrix term)

Where

PHIDh = log reading in 100% hydrocarbon

PHIDi = log reading in 100% of the ith component of matrix rock

PHID = log reading

PHIDsh = log reading in 100% shale PHIDw = log reading in 100% water

PHIe = effective porosity (fractional)

Sxo = water saturation in invaded zone (fractional)

Vi = volume of ith component of matrix rock

Vsh = volume of shale (fractional)

To solve for porosity from the density log, we assume PHIDh, PHIDi, PHIDsh, PHIDw, and Vsh are known. We also assume PHIDw = PHIDh and Sxo = 1.0 when no gas is present. If gas is indicated, we make assumptions about PHIDh and Sxo, usually in the form of a correction factor to the gas free case, as described later. This response equation is rigorous. Mineral and fluid parameters are shown HERE. Shale properties are selected from the log in an obvious shale zone.

PHIdc - Porosity From the Density Log

Calculate density porosity from density data.

2: PHID = (DENS – DENSMA) / (DENSW – DENSMA)

Apply density shale correction:

3: PHIDSH = (DENSH – DENSMA) / (DENSW – DENSMA)

4: PHIdc = PHID – Vsh * PHIDSH

Apply density gas correction.

5: IF DENSITYGASSWITCH$ = "ON"

6: THEN PHIdc = KD * PHIdc

Where

DENS = density log reading in zone of interest (gm/cc or Kg/m3)

DENSMA = density log reading in 100% matrix rock (gm/cc or Kg/m3)

DENSSH = density log reading in 100% shale (gm/cc or Kg/m3)

DENSW = density log reading in 100% water (gm/cc or Kg/m3)

KD = density log gas correction (fractional)

PHID = porosity from uncorrected density log (fractional)

PHIdc = porosity from density log corrected for shale (fractional)

PHIDSH = apparent density log porosity of 100% shale (fractional)

Vsh = shale volume (fractional)

The density log corrected for shale is a very good approximation to porosity, but the log was not common before 1965, so sonic or neutron methods may be necessary for wells drilled before that time.

KD is in the range of 0.5 – 1.0 depending on invasion, gas density and local experience. A correction is almost always needed if gas is present. Use gas correction only if PHIdc is too high compared to other sources and if gas is known to be present. This correction may be necessary even in shaly sands, since the depth of investigation of the density log is deep enough to see beyond the flushed zone.

When Density Log is in Porosity Units, Porosity May Be Wrong Because The Porosity Scale Does Not Match The Rock Mineralogy. If density porosity data is in percent, rather than fractional, divide the data values by 100 before Step 2 and 3 are applied. No compaction correction is made to density log data.

Numerical Example

1. Assume a zone with:

   DENS = 2.15 gm/cc

   DENSW = 1.00 gm/cc

   DENSMA = 2.65 gm/cc

   Vsh = 0.33

   DENSSH = 2.60 gm/cc

   PHID = (2.15 – 2.65) / (1.00 - 2.65) = 0.30

   PHIDSH = (2.60 – 2.65) / (1.00 – 2.65) = 0.03

   PHIdc = 0.30 – 0.33 * 0.03 = 0.29

No gas correction is required.

POROSITY FROM DENSITY LOG WITH MATRIX OFFSET

One step that is often required is to convert apparent porosity on the density log into density units, then reconstitute porosity from this value corrected for a desired matrix and fluid value.

PHIDm – Translate Density Porosity to New Matrix and Fluid

Calculate density from density porosity.

7: DENS = (PHID * KD1 + (1 – PHID) * KD2

WHERE:

	English Units	*Metric Units*	*Log Scale*
KD1	1.00	1000	All
KD2	2.65	2650	Sandstone
KD2	2.71	2710	Limestone
KD2	2.87	2870	Dolomite

Calculate shale density.

8: DENSSH = (PHIDSH * KD1 + (1 – PHIDSH) * KD2

Calculate porosity with new matrix and fluid.

9: PHIDm = (DENS – DENSMA) / (DENSW – DENSMA)

10: PHIDSHm = (DENSSH – DENSMA) / (DENSW – DENSMA)

11: PHIdc = PHIDm – Vsh * PHIDSHm

Apply density gas correction.

12: IF DENSITYGASSWITCH$ = "ON"

13: THEN PHIdc = KD * PHIdc

Where

DENSSH = density log reading in 100% shale reconstituted from density porosity data (gm/cc or Kg/m3)

DENS = density value reconstituted from density porosity data (gm/cc or Kg/m3)

DENSMA = matrix density (gm/cc or Kg/m3)

DENSW = fluid density (gm/cc or Kg/m3)

PHID = porosity from uncorrected density log (fractional)

PHIdc = porosity from density log corrected for shale (fractional)

PHIDm = density porosity log reading corrected for matrix offset (fractional)

PHIDSH = density porosity log reading in 100% shale (fractional)

PHIDSHm = density porosity log reading in 100% shale corrected for matrix offset (fractional)

Vsh = volume of shale (fractional)

Well Logging

Well logging, also known as borehole logging is the practice of making a detailed record (a *well log*) of the geologic formations penetrated by a borehole. The log may be based either on visual inspection of samples brought to the surface (*geological* logs) or on physical measurements made by instruments lowered into the hole (*geophysical* logs).

Well logging can be done during any phase of a well's history; drilling, completing, producing and abandoning. Well logging is performed in boreholes drilled for the oil and gas, groundwater, mineral and geothermal exploration, as well as part of environmental and geotechnical studies.

Wireline Logging

The oil and gas industry uses wireline logging to obtain a continuous record of a formation's rock properties. These can then be used to infer further properties, such as hydrocarbon saturation and formation pressure, and to make further drilling and production decisions. Wireline logging is performed by lowering a 'logging tool' on the end of a wireline into an oil well (or borehole) and recording petrophysical properties using a variety of sensors. Logging tools developed over the years measure the electrical, acoustic, radioactive, electromagnetic, nuclear magnetic resonance, and other properties of the rocks and their contained fluids. For this article, they are broadly broken down by the main property that they respond to.

The data itself is recorded either at surface (real-time mode), or in the hole (memory mode)to an electronic data format and then either a printed record or electronic presentation called a "well log" is provided to the client, along with an electronic copy of the raw data. Well logging operations can either be performed during the drilling process, to provide real-time information about the formations being penetrated by the borehole, or once the well has reached Total Depth and the whole depth of the borehole can be logged.

There are many types of wireline logs and they can be categorized either by their function or by the technology that they use. "Open hole logs" are run before the oil or gas well is lined with pipe or cased. "Cased hole logs" are run after the well is lined with casing or production pipe.

Wireline logs can also be divided based on what physical properties they measure.

History

Conrad and Marcel Schlumberger, who founded Schlumberger Limited in 1926, are considered the inventors of electric well logging. Conrad developed the Schlumberger array, which was a technique for prospecting for metal ore deposits, and the brothers adopted that surface technique to subsurface applications. On September 5, 1927, a crew working for Schlumberger lowered an electric sonde or tool down a well in Pechelbronn, Alsace, France creating the first well log. In modern terms, the first log was a resistivity log that could be described as 3.5-metre upside-down lateral log.

In 1931, Henri George Doll and G. Dechatre, working for Schlumberger, discovered that the galvanometer wiggled even when no current was being passed through the logging cables down in the well. This led to the discovery of the spontaneous potential (SP) which was as important as the ability to measure resistivity. The SP effect was produced naturally by the borehole mud at the boundaries of permeable beds. By simultaneously recording SP and resistivity, loggers could distinguish between permeable oil-bearing beds and impermeable nonproducing beds.

In 1940, Schlumberger invented the spontaneous potential dipmeter; this instrument allowed the calculation of the dip and direction of the dip of a layer. The basic dipmeter was later enhanced by the resistivity dipmeter (1947) and the continuous resistivity dipmeter (1952). Oil-based mud (OBM) was first used in Rangely Field, Colorado in 1948. Normal electric logs require a conductive or water-based mud, but OBMs are nonconductive. The solution to this problem was the induction log, developed in the late 1940s.

The introduction of the transistor and integrated circuits in the 1960s made electric logs vastly more reliable. Computerization allowed much faster log processing, and dramatically expanded log data-gathering capacity. The 1970s brought more logs and computers. These included combo type logs where resistivity logs and porosity logs were recorded in one pass in the borehole.

The two types of porosity logs (acoustic logs and nuclear logs) date originally from the 1940s. Sonic logs grew out of technology developed during World War II. Nuclear logging has supplemented acoustic logging, but acoustic or sonic logs are still run on some combination logging tools.

Nuclear logging was initially developed to measure the natural gamma radiation emitted by underground formations. However, the industry quickly moved to logs that actively bombard rocks with nuclear particles. The gamma ray log, measuring the natural radioactivity, was introduced by Well Surveys Inc. in 1939, and the WSI neutron log came in 1941. The gamma ray log is particularly useful as shale beds which often provide a relatively low permeability cap over hydrocarbon reservoirs usually display a higher level of gamma radiation. These logs were important because they can be used in cased wells (wells with production casing). WSI quickly became part of Lane-Wells. During World War II, the US Government gave a near wartime monopoly on open-hole logging to Schlumberger, and a monopoly on cased-hole logging to Lane-Wells. Nuclear logs continued to evolve after the war.

The nuclear magnetic resonance log was developed in 1958 by Borg Warner. Initially the NMR log was a scientific success but an engineering failure. However, the development of a continuous NMR logging tool by Numar (now a subsidiary of Halliburton) is a promising new technology.

Many modern oil and gas wells are drilled directionally. At first, loggers had to run their tools somehow attached to the drill pipe if the well was not vertical. Modern techniques now permit continuous information at the surface. This is known as logging while drilling (LWD) or measurement-while-drilling (MWD). MWD logs use mud pulse technology to transmit data from the tools on the bottom of the drillstring to the processors at the surface.

Electrical Logs

Resistivity: Resistivity logging works by characterizing the rock or sediment in a borehole by measuring its electrical resistivity.

Image Log

Image logs use a rotating transducer to measure acoustic impedance across the entire borehole wall. This can then be used to identify the presence and direction of rock fractures, as well as understanding the dip direction of the stratigraphy.

Porosity Logs

Porosity logs measure the fraction or percentage of pore volume in a volume of rock. Most porosity logs use either acoustic or nuclear technology. Acoustic logs measure characteristics of sound waves propagated through the well-bore environment. Nuclear logs utilize nuclear reactions that take place in the downhole logging instrument or in the formation. Nuclear logs include density logs and neutron logs, as well as gamma ray logs which are used for correlation.

The basic principle behind the use of nuclear technology is that a neutron source placed near the formation of which the porosity is required to be measured will result in neutrons being scattered by the hydrogen atoms, largely those present in the formation fluid. Since there is little difference in the neutrons scattered by hydrocarbons or water, the porosity measured gives a figure close to the true physical porosity whereas the figure obtained from electrical resistivity measurements is that due to the conductive formation fluid. The difference between neutron porosity and electrical porosity measurements therefore indicates the presence of hydrocarbons in the formation fluid.

Density

The density log measures the bulk density of a formation by bombarding it with a radioactive source and measuring the resulting gamma ray count after the effects of Compton Scattering and Photoelectric absorption. This bulk density can then be used to determine porosity.

Neutron Porosity

The neutron porosity log works by bombarding a formation with high energy epithermal neutrons that lose energy through elastic scattering to near thermal levels before being absorbed by the nuclei of the formation atoms. Depending on the particular type of neutron logging tool, either the gamma ray of capture, scattered thermal neutrons or scattered, higher energy epithermal neutrons are detected. The neutron porosity log is predominantly sensitive to the quantity of hydrogen atoms in a particular formation, which generally corresponds to rock porosity.

Boron is known to cause anomalously low neutron tool count rates due to it having a high capture cross section for thermal neutron absorption. An increase in hydrogen concentration in clay minerals has a similar effect on the count rate.

Sonic

A sonic log provides a formation interval transit time, which typically varies lithology and rock texture but particularly porosity. The logging tool consists of a piezoelectric transmitter and receiver and the time taken to for the sound wave to travel the fixed distance between the two is recorded as an *interval transit time.*

Lithology Logs

Gamma Ray: A log of the natural radioactivity of the formation along the borehole, measured in API, particularly useful for distinguishing between sands and shales in a siliclastic environment. This is because sandstones are usually nonradioactive quartz, whereas shales are naturally radioactive due to potassium isotopes in clays, and adsorbed uranium and thorium.

Self/Spontaneous Potential

The Spontaneous Potential (SP) log measures the natural or spontaneous potential difference between the borehole and the surface, without any applied current. It was one of the first wireline logs to be developed, found when a single potential electrode was lowered into a well and a potential was measured relative to a fixed reference electrode at the surface.

The most useful component of this potential difference is the electrochemical potential because it can cause a significant deflection in the SP response opposite permeable beds. The magnitude of this deflection depends mainly on the salinity contrast between the drilling mud and the formation water, and the clay content of the permeable bed. Therefore the SP log is commonly used to detect permeable beds and to estimate clay content and formation water salinity.

Miscellaneous

Caliper: A tool that measures the diameter of the borehole, using either 2 or 4 arms. It can be used to detect regions where the borehole walls are compromised and the well logs may be less reliable.

Nuclear Magnetic Resonance

Nuclear magnetic resonance (NMR) logging uses the NMR response of a formation to directly determine its porosity and permeability, providing a continuous record along the length of the borehole.

Logging While Drilling (LWD)

In the 1970s, a new approach to wireline logging was introduced in the form of logging while drilling (LWD). This technique provides similar well information to conventional wireline logging but instead of sensors being lowered into the well at the end of wireline cable, the sensors are integrated into the drill string and the measurements are made in real-time, whilst the well is being drilled. This allows drilling engineers and geologists to quickly obtain information such as porosity, resistivity, hole direction and weight-on-bit and they can use this information to make immediate decisions about the future of the well and the direction of drilling.

In LWD, measured data is transmitted to the surface in real time via pressure pulses in the well's mud fluid column. This mud telemetry method provides a bandwidth of less than 10 bits per second, although, as drilling through rock is a fairly slow process, data compression techniques mean that this is an ample bandwidth for real-time delivery of information. A higher sample rate of data is recorded into memory and retrieved when the drillstring is withdrawn at bit changes. High-definition downhole and subsurface information is available through networked or wired drillpipe that deliver memory quality data in real time.

Memory Log

This method of data acquisition involves recording the sensor data into a down hole memory, rather than transmitting "Real Time" to surface. There are some advantages and disadvantages to this memory option.

- The tools can be conveyed into wells where the trajectory is deviated or extended beyond the reach of conventional Electric Wireline cables. This can involve a combination of weight to strength ratio of the electric cable over this extended reach. In such cases the memory tools can be conveyed on Pipe or Coil Tubing.
- The type of sensors are limited in comparison to those used on Electric Line, and tend to be focussed on the cased hole,production stage of the well. Although there are now developed some memory "Open Hole" compact formation evaluation tool combinations. These tools can be deployed and carried downhole concealed internally in drill pipe to protect them from damage while running in the hole, and then "Pumped" out the end at depth to initiate logging. Other basic open hole formation evaluation memory tools are available for use in "Commodity" markets on slickline to reduce costs and operating time.
- In cased hole operation there is normally a "Slick Line" intervention unit. This uses a solid mechanical wire (.82 - .125 inches in OD), to manipulate or otherwise carry out operations in the well bore completion system. Memory operations are often carried out on this Slickline conveyance in preference to mobilizing a full service Electric Wireline unit.
- Since the results are not known until returned to surface, any realtime well dynamic changes cannot be monitored real time. This limits the ability to modify or change the well down hole production conditions accurately during the memory logging by changing the surface production rates. Something that is often done in Electric Line operations.
- Failure during recording is not known until the memory tools are retrieved. This loss of data can be a major issue on large offshore (expensive) locations. On land locations (e.g. South Texas, US) where there is what is called a "Commodity" Oil service sector, where logging often is without the rig infrastructure. This is less problematic, and logs are often run again without issue.

Coring

Figure: *An example of a granite core*

Coring is the process of obtaining an actual sample of a rock formation from the borehole. There are two main types of coring: 'full coring', in which a sample of rock is obtained using a specialised drill-bit as the borehole is first penetrating the formation and 'sidewall coring', in which multiple samples are obtained from the side of the borehole after it has penetrated through a formation. The main advantagea of sidewall coring over full coring are that it is cheaper (drilling doesn't have to be stopped) and multiple samples can be easily acquired, with the main disadvantages being that there can be uncertainty in the depth at which the sample was acquired and the tool can fail to acquire the sample.

Mudlogging

Mud logs are well logs prepared by describing rock or soil cuttings brought to the surface by mud circulating in the borehole. In the oil industry they are usually prepared by a mud logging company contracted by the operating company. One parameter a typical mud log displays is the formation gas (gas units or ppm). "The gas recorder usually is scaled in terms of arbitrary gas units, which are defined differently by the various gas-detector manufactures. In practice, significance is placed only on relative changes in the gas concentrations detected." The current oil industry standard mud log normally includes real-time drilling parameters such as rate of penetration (ROP), lithology, gas hydrocarbons, flow line temperature (temperature of the drilling fluid) and chlorides but may also include mud weight, estimated pore pressure and corrected d-exponent (corrected drilling exponent) for a pressure pack log. Other information that is normally notated on a mud log include directional data (deviation surveys), weight on bit, rotary speed, pump pressure, pump rate, viscosity, drill bit info, casing shoe depths, formation tops, mud pump info, to name just a few.

Information Use

In the oil industry, the well and mud logs are usually transferred in 'real time' to the operating company, which uses these logs to make operational decisions about the well, to correlate formation depths with surrounding wells, and to make interpretations about the quantity and quality of hydrocarbons present. Specialists involved in well log interpretation are called log analysts.

Well Test Objectives

Reservoir Evaluation

Reservoir evaluation is the evaluation of reservoir properties from log, core and pressure transient data. Including the interpretation of openhole well logs and pressure drawdown and buildup tests. This interpretation is usually carried out by Petrophysicists in close collaboration with Geologists and Reservoir engineers. It is a branch of reservoir engineering which is a subset of petroleum engineering.

Reservoir Characteristics

Broadly speaking, there are three main reservoir characteristics that matter to production. The character of the reservoir rock, the composition and purity of the crude, and the strength and nature of the drive mechanism all influence the flow rate and ultimate productivity of a reservoir. Reservoir depth, orientation, and complexity are also important in that these factors impact the cost and complexity of drilling, as does the composition of the cap rock and other matrix through which drilling must occur.

Reservoir Rock Characteristics

Though often referred to as underground pools of oil, reservoirs should not be conceptualized as large voids filled with crude. A better description is that reservoirs are subterranean expanses of rock whose pore spaces and fractures are saturated with some combination of oil, water, and other elements and molecules. It is better to think of reservoirs as rock sponges rather than underground milk jugs.

The two main characteristics of reservoir rocks that matter most to production costs, and ultimately to production rates, are porosity and permeability. *Porosity* and *permeability* are the two most important of reservoir rock characteristics. Together they influence the amount of trapped hydrocarbons and the rate at which the hydrocarbons will flow during primary production. The pore space in highly porous reservoir rocks allows them to hold more hydrocarbons (and also water), while the permeability of reservoir rocks impacts flow rates. The vast majority of reservoir rocks are sedimentary rocks, but highly fractured igneous and metamorphic rocks sometimes contain substantial reservoirs as well.

Porosity is expressed as a percentage and is calculated by dividing pore volume by total volume. Porosity depends on *grain shape* and and to a lesser extent on *grain sorting*. The theoretical upper bound of porosity is 48% for perfectly spherical grains of equal size. Evenly eroded particles will be more spherical in shape, while less eroded particles are more angular.

The 'porosity cutoff' is defined as the minimum porosity considered viable from an economic perspective, and porosity cutoff values vary by rock type. Typically, the porosity cutoff for sandstones is 8-10%, and the porosity cutoff for limestones is 3-5%. Limestone's lower porosity cutoff values reflect the propensity for limestones to be highly fractured. In addition, the porosity cutoff depends both on the depth and the "economics of the well" – the costs of production and market price for that particular crude stream (Hyne p. 156). As the price of crude rises, it makes economic sense to produce from less porous, or 'tight', reservoir rocks. In order to produce 'tight' formations, the rock must first be fractured, and this process is both energy intensive and costly.

Absolute permeability is measured in Darcies or millidarcies, and is often referred to as *perm* or *perm K*. Technically, a darcy is "the permeability that will allow a flow of 1 cubic centimeter per second of a fluid with 1 centipoise viscosity (resistance to flow) through a distance of 1 centimeter through an area of 1 square centimeter under a differential pressure of 1 atmosphere" (Hyne p. 157). From this definition, we see that low viscosity hydrocarbons (those with few carbons but a high carbon to hydrogen ratio) will flow at higher rates than the "barbed wire" asphaltenes. As a consequence, we can also deduce that in complex crude streams, early production will favour the lighter hydrocarbons and later production will be more highly comprised of heavy crudes – a very important factor to keep in mind as we examine models that forecast production rates at the global aggregate level.

While *absolute permeability* concerns a single fluid moving through rock, relative permeability concerns the relative movement of a fluid (hydrocarbon) through rock when another fluid (typically briny water) is also present. Relative permeability ranges from 1.0 (when there is only on fluid) to zero. The relative permeability depends on the ratio of oil to water, but it also depends on the absolute difference in viscosity between the crude stream and the brine. The relative permeability of heavy crude streams is decidedly less than that of light crude streams. From this fact we can deduce that as a reservoir ages, that is to say as production from a single reservoir progresses over time, not only will the crude stream become more viscous, the watercut (percentage of water produced with the oil) will increase.

The permeability of a source rock is primarily a function of two variables: grain shape, and grain sorting. Because these are the same variables that influence porosity, we see that permeability is strongly correlated to porosity, *but* highly porous rocks can have little or no permeability if the *pore throat* (the constriction between grains) is small. Conglomerate source rocks have low permeability, while well worn and sorted sandstones have high permeability. While rocks with very low porosity are by definition highly impervious, hydrocarbons can flow relatively quickly through the fractures in highly fractured rock.

Sandstones, limestones, and dolomites are the most productive reservoir rocks, and within group variation of porosity and permeability is greater than between group variation. Roughly half of all producing source rocks are sandstones. Limestones and dolomites make up nearly all of the difference (Hyne 2001). Because the volume of dolomite is less than that of calcite, the replacement of calcite by dolomite increases the pore space by 13% (Schlumberger online oilfield glossary). In addition, dolomites, like limestone, have a propensity to be highly fractured. Hence, dolomites make some of the most productive reservoir rocks. Conglomerates, igneous rocks, and metamorphic rocks *can* also produce at high rates provided they are highly fractured.

At the opposite end of the spectrum, siltstones and shales can be hydrocarbon rich, but because they are very low porosity and have a perm K values near zero, the hydrocarbons are trapped in place. Unless they are fractured (either through natural processes of folding or through mechanical processes like hydro-fracing) shales and siltstones will not flow.

Composition of Crude

A wide mix of hydrocarbon types, from methane to asphaltenes, are found in every crude stream or reservoir (a single subsurface accumulation of crude which is physically separated from other reservoirs and has a single natural pressure system). As Hyne puts it, in addition to the hydrocarbon molecules, "all sorts of cats and dogs can turn up in a hydrocarbon reservoir, whether it contains oil or gas or both. Along with natural gas and its natural gas liquid constituents, other gases can present themselves.

Carbon dioxide (CO_2), oxygen (O_2), nitrogen (N_2), and hydrogen sulfide (H_2S) are the most common" (Hyne 2001, p. 65). In addition to these gases, reservoirs also contain metals, salts, sulphur, and nitrogen (elements which often attach to the hydrocarbons themselves). For this reason, every reservoir is unique, and crude streams are typically characterized and priced by a set of easily measured characteristics: specific gravity, viscosity, sulphur content (often described as 'sweet' or 'sour'), nitrogen content, pour point, and flash point.

The weight, or specific gravity, of a crude is expressed in °API (read: degrees API). The arcane formula for generating the °API value reveals this measure to be inversely related to specific gravity. The higher the specific gravity, the lower the °API.

$$°API = (141.5 \chi \text{ specific gravity}) - 131.5$$

Using this formula, we can calculate the °API of water (specific gravity = 1) to be 10. Crude streams which have a higher specific gravity than water, like bitumen (a common term for asphaltic crudes), will have a °API value less than 10. Very light streams have higher values (sometimes in excess of 50).

The weight of a crude gives an indication of its hydrocarbon composition. Light crude streams tend to have a larger number of liquid paraffins, while heavy crude streams have a much higher asphaltic content. Within the industry, five classifications are generally employed to describe a crude stream.

In order from heaviest to lightest, we have bitumen (<10 °API), heavy crude oil (10° to 20° API), medium crude oil (20° to 35° API), light crude oil (35° to 50° API), and condensate (°50+ API) (Raymond and Leffler 2006).

Light crude streams "almost always command a higher price than heavier crudes" because they produce higher cuts of gasoline through fractional distillation, and, at least here in the U.S. where most cars run on gasoline rather than diesel, the gasoline cut offers the highest crack spread (the difference in price between equal volumes of crude and petroleum products) (Raymond and Leffler 2006, p. 64).

Heavy crudes are also highly viscous. They do not flow as easily, hence heavy crudes are more costly to move once they are produced. High viscosity crudes also challenge production as well because large, complex hydrocarbons have a much more difficult time squeezing through the *pore throats* of reservoirs. As heavy crudes are heated, the viscosity is lowered. You can see this by heating olive oil into a cold pan. As the oil heats up, it's viscosity decreases. Heavy crudes must sometimes be either heated or mixed with lighter crudes in order to flow under normal atmospheric conditions. This is especially in cold climates.

Waxy crudes have a relatively high pour point – the lowest temperature at which a crude stream will still pour before it solidifies. Pour points vary between 125°F and -75°F (Hyne 2001). Like heavy crudes, waxy crudes with very high pour points must be treated in order to flow easily under normal atmospheric conditions.

The sulphur content of crude streams is also measured as a percentage of ???. Crude streams with a high sulphur content, are colloquially referred to as 'sour crude' while those with a low sulphur content are referred to as 'sweet crude'. The etymology of sour and sweet modifiers is indeed derived from their flavour. In an earlier era, when the most value petroleum product was kerosene (i.e. lamp oil), products with a high sulphur content would create a non-odoriferous scent when burned. The best test of sulphur content that a purchaser could make was to literally taste the kerosene. If a sour flavour was detected, the sulphur content of the kerosene was high.

What the public did not know at the time was that the off-putting scent was sulphuric acid, a long-recognized environmental pollutant. Through the combustion process, sulphur combines with oxygen creating sulphur dioxide (SO_2). When sulphur dioxide is released into the atmosphere, it combines with water (H_2O) and produces sulphuric acid which materializes as acid rain. Recognizing this effect, the sulphur content of refined petroleum products is regulated, and, therefore, sulphur must be removed from crude.

For various reasons, sulphur content of crude has increased over time. Produced sulphur must be separated from the hydrocarbons which leaves the problem of what to do with all that sulphur. One idea is to form the sulphur into giant blocks, and use these sulphur blocks to build modern day pyramids like the ones that are currently being erected in Canada.

Carbon dioxide (CO_2) is often dissolved in the produced hydrocarbons. In addition to being a greenhouse gas, carbon dioxide is also an agent of corrosion (source ???). For this reason, carbon dioxide is removed from the crude stream at the site of production, liquefied, and reinjected into the reservoir.

Water-Oil Ratios, Watercut, and Relative Permeability

Water is present in nearly every hydrocarbon reservoir, and as a consequence, water is produced along with oil in nearly every well. The *water-oil ratio* influences production in five important ways.

First, salt content of the briny water accelerates the corrosion of all the equipment that it comes into contact with.

Second, the water which is produced must be separated from the oil. In cases where the water and oil are emulsified, the separation occurs in eumlusion heater treaters. When water is not held in suspension, it separated in a horizontal free water knockout separator.

The third important way that watercut influences the cost of production is that the water produced contains environmental contaminants (like heavy metals and sulphur). The water must either be re-injected or treated. In some cases, re-injection is not feasible, and the producer must treat the water.

The fourth important way that watercut can impact production costs occurs in wells that are low pressure and high temperature – conditions which are "ripe for free water and natural gas to form hydrates, a slushy mass that can plug flow lines... Remedies include keeping the gas warm – above the hydrate dew point" (Raymond and Leffler, p. 192).

The fifth important way in which water influences production is in the production rates themselves. When watercuts – the ratio of water production to total production – are high, the total volume of crude produced will be far less than the quantity of total liquids flowing through the wellhead. Over time, as the watercut increases, a constant flow rate of total liquids will bear witness to declining production rates of crude.

Decline Curve Analysis

Decline curve analysis is a means of predicting future oil well or gas well production based on past production history. Production decline curve analysis is a traditional means of identifying well production problems and predicting well performance and life based on measured oil well production.

Before the availability of computers, decline curve analysis was performed by hand on semi-log plot paper. Currently, decline curve analysis software on PC computers is used to plot production decline curves for petroleum economics analysis.

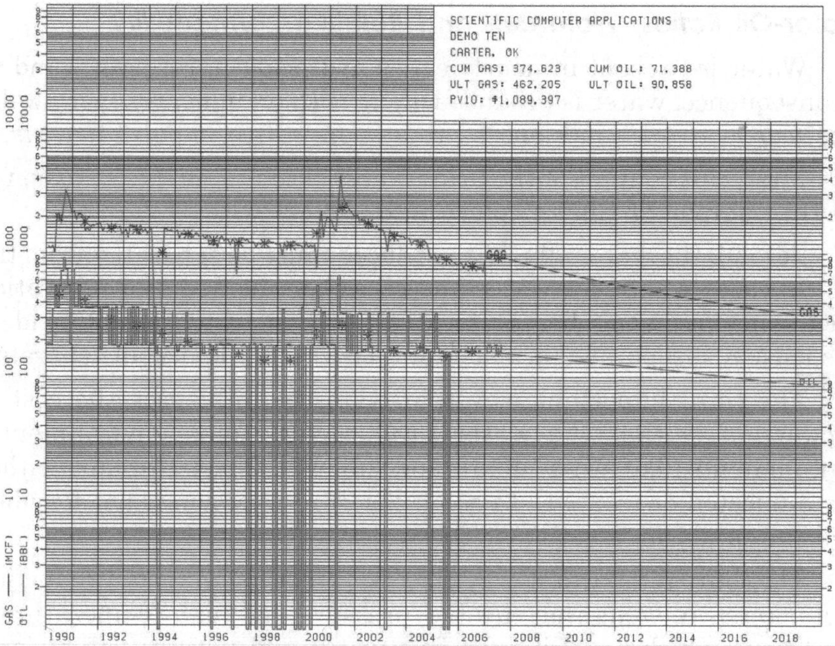

Figure: *Decline curve generated by decline curve analysis software, utilized in petroleum economics to indicate the depletion of oil & gas in a petroleum reservoir.*

Background

Oil and gas wells usually reach their maximum output shortly after completion. From that time, other than wells completed in water-drive reservoirs, they decline in production, the rapidity of decline depending on the output of the wells and on other factors governing their productivity. The production decline curve shows the amount of oil and gas produced per unit of time for several consecutive periods; if the conditions affecting the rate of production are not changed, the curve may be fairly regular, and, if projected, will furnish useful knowledge as to the future production of the well. By the aid of this knowledge the value of a property may be judged, and proper depletion and depreciation charges may be made on the books of the operating company.

Production decline curve analysis is important in determining the value in oil and gas wells in oil and gas economics. Decline curves are the most common means of forecasting oil and gas production. Decline curves have many advantages: they use data which is easy to obtain, they are easy to plot, they yield results on a time basis, and they are easy to analyze. Decline curves are also one of the oldest methods of predicting oil reserves.

Chapter 7
Seismic Interferometry

Interferometry examines the general interference phenomena between pairs of signals in order to gain useful information about the subsurface. Seismic interferometry (SI) utilizes the crosscorrelation of signal pairs to reconstruct the impulse response of a given media. Jon Claerbout's initial (1968) conjecture provided the framework upon which modern theory is based.

A signal at a location A can be crosscorrelated with a signal at a location B to reproduce a virtual source-receiver pair using seismic interferometry. Crosscorrelation is often considered the key mathematical operation in this approach, but it is also possible to use convolution to come up with a similar result. The crosscorrelation of passive noise measured at a free surface reproduces the subsurface impulse response. As such, it is possible to obtain information about the subsurface with no need for an active seismic source. This method, however, is not limited to passive sources, and can be extended for use with active sources and computer–generated waveforms.

As of 2006 the field of seismic interferometry was beginning to change the way geophysicists view noise. Seismic interferometry uses this previously–ignored noise in models of the shallow subsurface. Potential applications include both research and industry.[1]

History and Development

Claerbout (1968) first considered seismic interferometry for investigating the shallow subsurface. He later predicted that seismic interferometry could be applied to real world media, which was later proven. The long term average of random ultrasound waves can reconstruct the impulse response between two points on an aluminium block. However, they had assumed random diffuse noise, limiting interferometry in real world conditions. In a similar case, it was shown that the expressions for uncorrelated noise sources reduce to a single crosscorrelation of observations at two receivers. The interferometric impulse response of the subsurface can be reconstructed using only an extended record of background noise, initially only for the surface and direct wave arrivals.

Crosscorrelations of seismic signals from both active and passive sources at the surface or in the subsurface can be used to reconstruct a valid model of the subsurface. Seismic interferometry can produce a result similar to traditional methods without limitations on the diffusivity of the wavefield or ambient sources. In a drilling application, it is possible to utilize a virtual source to image the subsurface adjacent to a downhole location. This application is increasingly utilized particularly for exploration in subsalt settings.

Mathematical and Physical Explanation

Seismic interferometry provides for the possibility of reconstructing the subsurface reflection response using the crosscorrelations of two seismic traces. Following Claerbout's conjecture, recent work has mathematically demonstrated applications of crosscorrelation for reconstructing Green's function using wave field reciprocity theorem in a lossless, 3D heterogeneous medium. Traces are most often extended records of passive background noise, but it is also possible to utilize active sources depending on the objective. Seismic interferometry essentially exploits the phase difference between adjacent receiver locations to image the subsurface.

Seismic interferometry consists of simple crosscorrelation and stacking of actual receiver responses to approximate the impulse response as if a virtual source was placed at the location of the applicable receiver. Crosscorrelation of continuous functions in the time domain is presented as Equation 1.

Equation 1

$$(f\,1 * f\,2)(t) = \int f\,1(\lambda) f\,2(\lambda - t)dt$$

Where the functions are integrated as a function of time at different lag values. In fact, crosscorrelation can be understood conceptually as the traveltime lag associated with waveforms in two discrete receiver locations. Crosscorrelation is similar to convolution where the second function is folded relative to the first.

Seismic interferometry is fundamentally similar to the optical interferogram produced by the interference of a direct and reflected wave passing through a glass lens where intensity is primarily dependent upon the phase component.

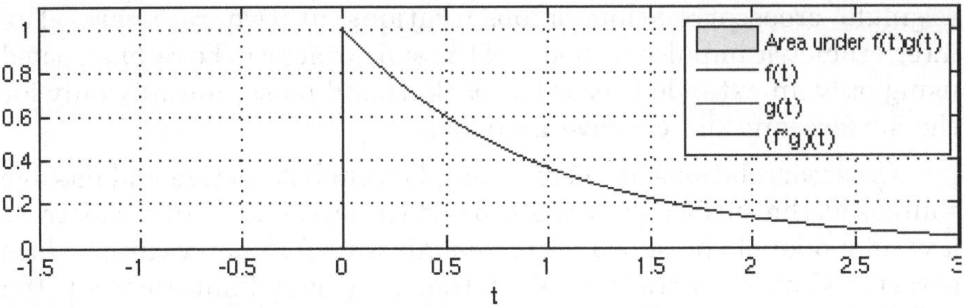

Figure: *Principle of convolution. Similar to cross-correlation*

Equation 2

$$I = 1+2R2\ \cos[\omega(\lambda Ar+\lambda rB)]+R\hat{}4$$

Where: Intensity is related to the magnitude of the reflection coefficient (R) and the phase component ω(λAr+λrB). An estimate of the reflectivity distributions can be obtained through the crosscorrelation of the direct wave at a location A with the reflection recorded at a location B where A represents the reference trace.

The multiplication of the conjugate of the trace spectrum at A and the trace spectrum at B gives:

Equation 3

$$\Phi AB = Re\hat{}i\omega(\lambda Ar+\lambda rB) + o.t.$$

Where: ΦAB = product spectrum o.t. = additional terms, e.g. correlations of directdirect, etc. As in the previous case, the product spectrum is a function of phase.

Key: Changes in reflector geometry lead to changes in the correlation result and the reflector geometry can be recovered through the application of a migration kernel.

Interpretation of raw interferograms is not normally attempted; crosscorrelated results are generally processed using some form of migration.

In the simplest case, consider a rotating drill bit at depth radiating energy that is recorded by geophones on the surface. It is possible to assume that the phase of the source wavelet at a given position is random and utilize the crosscorrelation of the direct wave at a location A with a ghost reflection at a location B to image a subsurface reflector without any knowledge regarding the source location.

The crosscorrelation of traces A and B in the frequency domain simplifies as:

Equation 4

$$\Phi\,(A, B) = -(Wi\omega)\text{^}2\ Re\text{^}i\omega(\lambda Ar\lambda rB)\text{+o.t.}$$

Where: Wi(ω) = frequency domain source wavelet (ith wavelet)

The crosscorrelation of the direct wave at a location A with a ghost reflection at a location B removes the unknown source term where:

Equation 5

$$\Phi\,(A,B)H{-}Re\text{^}i\omega(\lambda Ar\lambda rB)$$

This form is equivalent to a virtual source configuration at a location A imaging hypothetical reflections at a location B. Migration of these correlation positions removes the phase term and yields a final migration image at position x where:

$m(x) = \Sigma\psi(A,B,\lambda Ax+\lambda xB)$ Where: $\psi(A,B,t)$ = temporal correlation between locations A and B with lag time t

This model has been applied to simulate subsurface geometry in West Texas using simulated models including a traditional buried source and a synthetic (virtual) rotating drill bit source to produce similar results. A similar model demonstrated the reconstruction of a simulated subsurface geometry.

In this case, the reconstructed subsurface response correctly modelled the relative positions of primaries and multiples. Additional equations can be derived to reconstruct signal geometries in a wide variety of cases.

Applications

Seismic interferometry is currently utilized primarily in research and academic settings. In one example, passive listening and the crosscorrelation of long noise traces was used to approximate the impulse response for shallow subsurface velocity analysis in Southern California. Seismic interferometry provided a result comparable to that indicated using elaborate inversion techniques. Seismic interferometry is most often used for the examination of the near surface and is often utilized to reconstruct surface and direct waves only.

As such, seismic interferometry is commonly used to estimate ground roll to aid in its removal. Seismic interferometry simplifies estimates of shear wave velocity and attenuation in a standing building. Seismic interferometry has been applied to image the seismic scattering and velocity structure of volcanoes.

Exploration and Production

Increasingly, seismic interferometry is finding a place in exploration and production. SI can image dipping sediments adjacent to salt domes. Complex salt geometries are poorly resolved using traditional seismic reflection techniques. An alternative method calls for the use of downhole sources and receivers adjacent to subsurface salt features. It is often difficult to generate an ideal seismic signal in a downhole location. Seismic interferometry can virtually move a source into a downhole location to better illuminate and capture steeply dipping sediments on the flank of a salt dome.

In this case, the SI result was very similar to that obtained using an actual downhole source. Seismic interferometry can locate the position of an unknown source and is often utilized in hydrofrac applications to map the extent of induced fractures. It is possible that interferometric techniques can be applied to timelapse seismic monitoring of subtle changes in reservoir properties in the subsurface.

Limitations

Seismic interferometry applications are currently limited by a number of factors. Real world media and noise represent limitations for current theoretical development. For example, for interferometry to work noise sources must be uncorrelated and completely surround the region of interest. In addition, attenuation and geometrical spreading are largely neglected and need to be incorporated into more robust models.

Other challenges are inherent to seismic interferometry. For example, the source term only drops out in the case of the crosscorrelation of a direct wave at a location A with a ghost reflection at a location B. The correlation of other waveforms can introduce multiples to the resulting interferogram. Velocity analysis and filtering can reduce but not eliminate the occurrence of multiples in a given dataset.

Although there have been many advancements in seismic interferometry challenges still remain. One of the biggest remaining challenges is extending the theory to account for real world media and noise distributions in the subsurface. Natural sources typically do not comply with mathematical generalizations and may in fact display some degree of correlation. Additional problems must be addressed before applications of seismic interferometry can become more widespread.

Hydraulic Fracturing Proppants

A proppant is a solid material, typically treated sand or man-made ceramic materials, designed to keep an induced hydraulic fracture open, during or following a fracturing treatment. It is added to a fracking fluid which may vary in composition depending on the type of fracturing used, and can be gel, foam or slickwater-based. In addition, there may be unconventional fracking fluids.

Fluids make tradeoffs in such material properties as viscosity, where more viscous fluids can carry more concentrated proppant; the energy or pressure demands to maintain a certain flux pump rate (flow velocity) that will conduct the proppant appropriately; pH, various rheological factors, among others. In addition, fluids may be used in low-volume well stimulation of high-permeability sandstone wells (20k to 80k gallons per well) to the high-volume operations such as shale gas and tight gas that use millions of gallons of water per well.

Conventional wisdom has often vacillated about the relative superiority of gel, foam and slickwater fluids with respect to each other, which is in turn related to proppant choice. For example, Zuber, Kuskraa and Sawyer (1988) found that gel-based fluids seemed to achieve the best results for coalbed methane operations, but as of 2012, slickwater treatments are more popular.

Ignoring proppant, slickwater fracturing fluids are mostly water, generally 99% or more by volume, but gel-based fluids can see polymers and surfactants comprising as much as 7 vol% , ignoring other additives. Other common additives include hydrochloric acid (low pH can etch certain rocks, dissolving limestone for instance), friction reducers, guar gum, biocides, emulsion breaker and emulsifiers.

Radioactive tracer isotopes are sometimes included in the hydrofracturing fluid to determine the injection profile and location of fractures created by hydraulic fracturing. Patents describe in detail how several tracers are typically used in the same well. Wells are hydraulically fractured in different stages. Tracers with different half-lives are used for each stage. Their half-lives range from 40.2 hours (Lanthanum-140) to 5.27 years (Cobalt-60). Amounts per injection of radionuclide are listed in The US Nuclear Regulatory Commission (NRC) guidelines. The NRC guidelines also list a wide range or radioactive materials in solid, liquid and gaseous forms that are used as field flood or enhanced oil and gas recovery study applications tracers used in single and multiple wells.

Except for diesel-based additive fracturing fluids, noted by the American Environmental Protection Agency to have a higher proportion of volatile organic compounds and carcinogenic BTEX, use of fracturing fluids in hydraulic fracturing operations was explicitly excluded from regulation under the American Clean Water Act in 2005, a legislative move that has since attracted controversy for being the product of special interests lobbying.

Proppant Permeability and Mesh Size

Proppants used should be permeable or permittive to gas under high pressures; the interstitial space between particles should be sufficiently large, yet have the mechanical strength to withstand closure stresses to hold fractures open after the fracturing pressure is withdrawn. Large mesh proppants have greater permeability than small mesh proppants at low closure stresses, but will mechanically fail (i.e. get crushed) and produce very fine particulates ("fines") at high closure stresses such that smaller-mesh proppants overtake large-mesh proppants in permeability after a certain threshold stress.

Though sand is a common proppant, untreated sand is prone to significant fines generation; fines generation is often measured in wt% of initial feed. A commercial newsletter from Momentive cites untreated sand fines production to be 23.9% compared with 8.2% for lightweight ceramic and 0.5% for their product. One way to maintain an ideal mesh size (i.e. permeability) while having sufficient strength is to choose proppants of sufficient strength; sand might be coated with resin, to form CRCS (Curable Resin Coated Sand) or PRCS (Pre-Cured Resin Coated Sands). In certain situations a different proppant material might be chosen altogether—popular alternatives include ceramics and sintered bauxite.

Proppant Weight and Strength

Increased strength often comes at a cost of increased density, which in turn demands higher flow rates, viscosities or pressures during fracturing, which translates to increased fracturing costs, both environmentally and economically. Lightweight proppants conversely are designed to be lighter than sand (~2.5 g/cc) and thus allow pumping at lower pressures or fluid velocities. Light proppants are less likely to settle. Porous materials can break the strength-density trend, or even afford greater gas permeability. Proppant geometry is also important; certain shapes or forms amplify stress on proppant particles making them especially vulnerable to crushing (a sharp discontinuity can classically allow infinite stresses in linear elastic materials).

Proppant Deposition and Post-treatment Behaviours

Proppant mesh size also impacts fracture length: proppants can be "bridged out" if the fracture width decreases to less than twice the size of the diameter of the proppant. As proppants are deposited in a fracture, proppants can resist further fluid flow or the flow of other proppants, inhibiting further growth of the fracture. In addition, closure stresses (once external fluid pressure is released) may cause proppants to reorganise or "squeeze out" proppants, even if no fines are generated, resulting in smaller effective width of the fracture and decreased permeability. Some companies try to cause weak bonding at rest between proppant particles in order to prevent such reorganisation. The modelling of fluid dynamics and rheology of fracturing fluid and its carried proppants is a subject of active research by the industry.

Proppant Costs

Though good proppant choice positively impacts output rate and overall ultimate recovery of a well, commercial proppants are also constrained by cost. Transport costs from supplier to site form a significant component of the cost of proppants.

Steam Injection (Oil Industry)

Steam injection is an increasingly common method of extracting heavy crude oil. It is considered an enhanced oil recovery (EOR) method and is the main type of thermal stimulation of oil reservoirs. There are several different forms of the technology, with the two main ones being Cyclic Steam Stimulation and Steam Flooding. Both are most commonly applied to oil reservoirs, which are relatively shallow and which contain crude oils which are very viscous at the temperature of the native underground formation. Steam injection is widely used in the San Joaquin Valley of California (USA), the Lake Maracaibo area of Venezuela, and the oil sands of northern Alberta (Canada).

Another contributing factor that enhances oil production during steam injection is related to near-wellbore cleanup. In this case, steam reduces the interfacial tension that ties paraffins and asphaltenes to the rock surfaces while steam distillation of crude oil light ends creates a small solvent bank that can miscibly remove trapped oil.

Cyclic Steam Stimulation (CSS)

This method, also known as the Huff and Puff method, consists of 3 stages: injection, soaking, and production. Steam is first injected into a well for a certain amount of time to heat the oil in the surrounding reservoir to a temperature at which it flows. After it is decided enough steam has been injected, the steam is usually left to "soak" for some time after (typically not more than a few days). Then oil is produced out of the same well, at first by natural flow (since the steam injection will have increased the reservoir pressure) and then by artificial lift. Production will decrease as the oil cools down, and once production reaches an economically determined level the steps are repeated again.

The process can be quite effective, especially in the first few cycles. However, it is typically only able to recover approximately 20% of the Original Oil in Place (OOIP), compared to steam flooding, which has been reported to recover over 50% of OOIP. It is quite common for wells to be produced in the cyclic steam manner for a few cycles before being put on a steam flooding regime with other wells.

The mechanism was accidentally discovered by Shell while it was doing a steam flood in Venezuela and one of its steam injectors blew out and ended up producing oil at much higher rates than a conventional production well in a similar environment.

The use of steam injection to recover heavy oil has been in use in the oil fields of California since the 1950s. The Cyclic Steam Stimulation or "huff-and-puff" method has been in use by Imperial Oil at Cold Lake since 1985 and is also used by Canadian Natural Resources at Primrose and Wolf Lake and by Shell Canada at Peace River. In this method, the well is put through cycles of steam injection, soak, and oil production. First, steam is injected into a well at a temperature of 300 to 340 degrees Celsius for a period of weeks to months; then, the well is allowed to sit for days to weeks to allow heat to soak into the formation; and, later, the hot oil is pumped out of the well for a period of weeks or months. Once the production rate falls off, the well is put through another cycle of injection, soak and production. This process is repeated until the cost of injecting steam becomes higher than the money made from producing oil. The CSS method has the advantage that recovery factors are around 20 to 25% and the disadvantage that the cost to inject steam is high.

Canadian Natural Resources use "employs cyclic steam or "huff and puff" technology to develop bitumen resources. This technology requires one well bore and the production consists of the injection and production phases. First steam is "injected for several weeks, mobilizing cold bitumen". Then the flow "on the injection well is reversed producing oil through the same injection well bore. The injection and production phases together comprise one cycle. "Steam is re-injected to begin a new cycle when oil production rates fall below a critical threshold due to the cooling of the reservoir." (CNRL 2013)

High Pressure Cyclic Steam Stimulation (HPCSS)

"Roughly 35 per cent of all *in situ* production in the Alberta oil sands uses a technique called High Pressure Cyclic Steam Stimulation (HPCSS), which cycles between two phases: first, steam is injected into an underground oilsands deposit to soften the bitumen; then, the resulting hot mixture of bitumen and steam (called a "bitumen emulsion") is pumped up to the surface. The process is then repeated multiple times."(Severson-Baker 2013) An Alberta Energy Regulator (AER) news release explained the difference between high pressure cyclic steam stimulation (HPCSS) and steam assisted gravity drainage (SAGD). "HPCSS has been used in oil recovery in Alberta for more than 30 years. The method involves injecting high-pressure steam into a reservoir over a prolonged period of time. As heat softens the bitumen and water dilutes and separates the bitumen from the sand, the pressure creates cracks and openings through which the bitumen can flow back into the steam-injector wells. HPCSS differs from steam assisted gravity drainage (SAGD) operations where steam is injected at lower pressures without fracturing the reservoir and uses gravity drainage as the primary recovery mechanism."(2013 AER) Canadian Natural Resources Limited's (CNRL) Primrose and Wolf Lake in situ oil sands project near Cold Lake, Alberta in the Clearwater Formation, operated by CNRL subsidiary Horizon Oil Sands, use the high pressure cyclic steam stimulation (HPCSS).(2013 AER)

Primrose Thermal in Situ Oilsands Project

Canadian Natural Resources Limited's one of the largest independent crude oil and natural gas producers in the world, with its head office in Calgary, with operations are focused in Western Canadian Sedimentary basin,(CNRL & 2012 1) the North Sea and offshoreWest Africa. CNRL"s Primrose field produced to "99,000 barrels per day"(CNRL & 2012 1) Synthetic Crude Oil (SCO) in 2012, c.100,000 to 107,000 barrels per day in 2013(Healing 2013) and expects to "add value for decades." Canadian Natural uses the Kirby South Steam Assisted Gravity Drainage (SAGD) extraction(CNRL & 2012 1) method including both the cyclical and steam-assisted gravity drainage (SAGD) technologies the Primrose and Wolf Lake - Thermal In Situ located approximately 55 km north of Bonnyville.(CNRL 2013) The Canadian

Natural Resources Ltd.'s "operations use an "in situ" or underground extraction technology called cyclic steam stimulation, which involves injecting thousands of gallons of superhot, high-pressure steam into deep underground reservoirs. This heats and liquefies the hard bitumen and creates cracks through which the bitumen flows and is then pumped to the surface.(Pullman and Lukacs 2013)(Canadian Association of Petroleum Producers (CAPP) 2009)

Steam Flooding

In a steam flood, sometimes known as a steam drive, some wells are used as steam injection wells and other wells are used for oil production. Two mechanisms are at work to improve the amount of oil recovered. The first is to heat the oil to higher temperatures and to thereby decrease its viscosity so that it more easily flows through the formation toward the producing wells. A second mechanism is the physical displacement employing in a manner similar to water flooding, in which oil is meant to be pushed to the production wells. While more steam is needed for this method than for the cyclic method, it is typically more effective at recovering a larger portion of the oil.

A form of steam flooding that has become popular in the Alberta oil sands is steam assisted gravity drainage (SAGD), in which two horizontal wells are drilled, one a few metres above the other, and steam is injected into the upper one. The intent is to reduce the viscosity of the bitumen to the point where gravity will pull it down into the producing well.

In 2011 Laricina Energy combined solvent injection with steam injection in a process called solvent cyclic steam-assisted gravity drainage (SC-SAGD).(Canadian Association of Petroleum Producers (CAPP) 2009) Laricina claims that combining solvents with steam reduces the overall steam oil ratio for recovery by 30%.

Gas Reinjection

Gas reinjection is the reinjection of natural gas into an underground reservoir, typically one already containing both natural gas and crude oil, in order to increase the pressure within the reservoir and thus induce the flow of crude oil or else sequester gas that cannot be exported. This is not to be confused with gas lift, where gas is injected into the annulus of the well rather than the reservoir. After the crude has been pumped out, the natural gas is once again recovered. Since many of the wells found around the world contain heavy crude, this process increases their production. The basic difference between light crude and heavy crude is its viscosity and pumpability - the lighter the crude the easier it is to pump. Recovery of hydrocarbons in a well is generally limited to 50% (heavy crudes) and 75-80% (light crudes). Recycling of natural gas or other inert gases causes the pressure to rise in the well, thus causing more gas molecules to dissolve in the oil lowering its viscosity and thereby increasing the well's output. Air is not suitable for repressuring wells because it tends to cause deterioration of the oil, thus carbon dioxide or natural gas is used to repressure the well. The term 'gas-reinjection' is also sometimes referred to as repressuring—the term being used only to imply that the pressure inside the well is being increased to aid recovery.

Injection or reinjection of carbon dioxide also takes place in order to reduce the emission of CO_2 into the atmosphere, a form of carbon sequestration. This has been proposed as a method to combat climate change, allowing mass storage of CO_2 over a geological timescale. Reinjection of carbon dioxide in the Norwegian Sleipner gas field saves the operators 1 million Norwegian Kroners per day in national carbon taxes.

Water Injection (Oil Production)

Water injection refers to the method in the oil industry where water is injected into the reservoir, usually to increase pressure and thereby stimulate production. Water injection wells can be found both on- and offshore, to increase oil recovery from an existing reservoir.

Water is injected (1) to support pressure of the reservoir (also known as voidage replacement), and (2) to sweep or displace oil from the reservoir, and push it towards a well.

Normally only 30% of the oil in a reservoir can be extracted, but water injection increases that percentage (known as the recovery factor) and maintains the production rate of a reservoir over a longer period.

Waterflooding began accidentally in Pithole, Pennsylvania by 1865. Waterflooding became common in Pennsylvania in the 1880s.

Sources of Injected Water

Any and every source of bulk water can be, and has been, used for injection. The following sources of water are used for recovery of oil:

Produced water is often used as an injection fluid. This reduces the potential of causing formation damage due to incompatible fluids, although the risk of scaling or corrosion in injection flowlines or tubing remains. Also, the produced water, being contaminated with hydrocarbons and solids, must be disposed of in some manner, and disposal to sea or river will require a certain level of clean-up of the water stream first. However, the processing required to render produced water fit for reinjection may be equally costly.

As the volumes of water being produced are never sufficient to replace all the production volumes (oil and gas, in addition to water), additional "make-up" water must be provided. Mixing waters from different sources exacerbates the risk of scaling.

Seawater is obviously the most convenient source for offshore production facilities, and it may be pumped inshore for use in land fields. Where possible, the water intake is placed at sufficient depth to reduce the concentration of algae; however, filtering, deoxygenation and biociding is generally required.

Aquifer water from water-bearing formations other than the oil reservoir, but in the same structure, has the advantage of purity where available.

River water will always require filtration and biociding before injection.

Filters

The filters must clean the water and remove any impurities, such as shells and algae. Typical filtration is to 2 micrometres, but really depends on reservoir requirements. The filters are so fine so as not to block the pores of the reservoir. Sand filters are a common used filtration technology to remove solid impurities from the water. The sand filter has different beds with various sizes of sand granules. The sea water traverses the first, coarsest, layer of sand down to the finest and to clean the filter, the process is inverted. After the water is filtered it continues on to fill the de-oxygenation tower. Sand filters are bulky, heavy, have some spill over of sand particles and require chemicals to enhance water quality. A more sophisticated approach is to use automatic selfcleaning backflushable screen filters (suction scanning) because these do not have the disadvantages sand filters have.

The importance of proper water treatment is often underestimated by oil companies and engineering companies. Especially with river-, and seawater, intake water quality can vary tremendously (algae blooming in spring time, storms and current stirring up sediments from the seafloor) which will have significant impact on the performance of the water treatment facilities. If not addressed correctly, water injection may not be successful. This results in poor water quality, clogging of the reservoir and loss of oil production.

De-oxygenation

Oxygen must be removed from the water because it promotes corrosion and growth of certain bacteria. Bacterial growth in the reservoir can produce toxic hydrogen sulfide, a source of serious production problems, and block the pores in the rock.

A deoxygenation tower brings the injection water into contact with a dry gas stream (gas is always readily available in the oilfield). The filtered water drops into the de-oxygenation tower, splashing onto a series of trays, causing dissolved oxygen to be lost to the gas stream.

An alternative method, also used as a backup to deoxygenation towers, is to add an oxygen scavenging agent such as sodium bisulfite and ammonium bisulphite.

Water Injection Pumps

The high pressure, high flow water injection pumps are placed near to the de-oxygenation tower and boosting pumps. They fill the bottom of the reservoir with the filtered water to push the oil towards the wells like a piston. The result of the injection is not quick, it needs time. Water injection is used to prevent low pressure in the reservoir. The water replaces the oil which has been taken, keeping the production rate and the pressure the same over the long term.

Gas Lift

Gas lift or Bubble pumps use the artificial lift technique of raising a fluid such as water or oil by introducing bubbles of compressed air, water vapor or other vaporous bubbles into the outlet tube. This has the effect of reducing the hydrostatic pressure in the outlet tube vs. the hydrostatic pressure at the inlet side of the tube.

Devices using this type of lift mechanism:

- Coffee percolators use vaporized water to lift hot water
- Airlift pumps uses compressed air to lift water
- Pulser pumps use a subterranean chamber of air for an airlift pump
- Suction dredges use an airlift pump to vacuum mud, sand and debris

- Mist lift pumps uses vaporized water to lift seawater in Ocean thermal energy conversion

Petroleum Industry Uses

In the United States, gas lift is used in 10% of the oil wells that have insufficient reservoir pressure to produce the well. In the petroleum industry, the process involves injecting gas through the tubing-casing annulus. Injected gas aerates the fluid to reduce its density; the formation pressure is then able to lift the oil column and forces the fluid out of the wellbore. Gas may be injected continuously or intermittently, depending on the producing characteristics of the well and the arrangement of the gas-lift equipment.

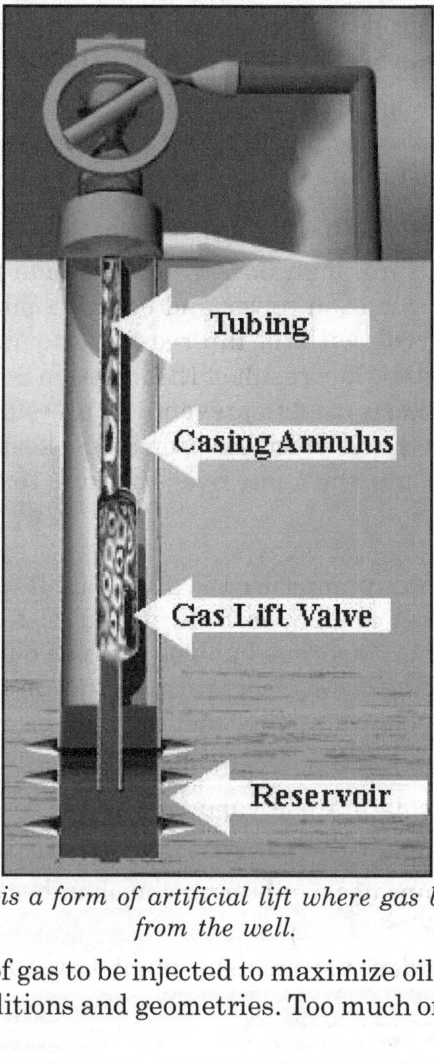

Figure: *Gas lift is a form of artificial lift where gas bubbles lift the oil from the well.*

The amount of gas to be injected to maximize oil production varies based on well conditions and geometries. Too much or too little injected gas will result in less than maximum production. Generally, the optimal amount of injected gas is determined by well tests, where the rate of injection is varied and liquid production (oil and perhaps water) is measured.

Although the gas is recovered from the oil at a later separation stage, the process requires energy to drive a compressor to raise the pressure of the gas to a level where it can be re-injected.

The gas-lift mandrel is a device installed in the tubing string of a gas-lift well onto which or into which a gas-lift valve is fitted. There are two common types of mandrels. In a conventional gas-lift mandrel, a gas-lift valve is installed as the tubing is placed in the well. Thus, to replace or repair the valve, the tubing string must be pulled. In the side-pocket mandrel, however, the valve is installed and removed by wireline while the mandrel is still in the well, eliminating the need to pull the tubing to repair or replace the valve.

A gas-lift valve is a device installed on (or in) a gas-lift mandrel, which in turn is put on the production tubing of a gas-lift well. Tubing and casing pressures cause the valve to open and close, thus allowing gas to be injected into the fluid in the tubing to cause the fluid to rise to the surface. In the lexicon of the industry, gas-lift mandrels are said to be "tubing retrievable" wherein they are deployed and retrieved attached to the production tubing.

History

Air lift uses compressed air to lift water in operations such as dredging and underwater archeology. It is also found in aquariums to keep water circulating. These forms of lift were used as far back as 1797 in mines to lift water from mine shafts. These systems used single point injection of air into the liquid stream, normally through a foot valve at the bottom of the string. Gas lift was used as early as 1864 in Pennsylvania to lift oil wells, also using compressed air, via an air pipe bringing the air to the bottom of the well. Air was used in Texas for large-scale artificial lift. In 1920 natural gas replaced air, lowering the risk of explosion. From 1929 until 1945 about 25000 patents were issued on different types of gas lift valves that could be used for unloading in stages. Some of these systems involved moving the tubing, or using wireline sinker bars to change the lift point. Others were spring operated valves. Ultimately, in 1944 W.R. King patented the pressurized bellows valve that is used today. In 1951 the sidepocket mandrel was developed for selectively positioning and retrieving gas lift valves with wireline.

Squeeze Job

Squeeze job, or squeeze cementing is a term often used in the oilfield to describe the process of injecting cement slurry into a zone, generally for pressure-isolation purposes. The term probably originated from the concept that enough water is "squeezed" out of the slurry to render it unflowable, so the portion that has actually entered the zone will stay in place when the squeeze pressure is removed. After surface indications (e.g., pressure reaching a predetermined maximum) that a squeeze has been attained, any still-pumpable cement slurry remaining in the drill pipe or tubing ideally can be reverse circulated out before it sets.

Usually the zone to be squeezed is isolated from above with a packer (and possibly from below with a bridge plug), but sometimes the squeezing pressure is applied to the entire casing string in what is known as a bradenhead squeeze, (named for an old manufacturer of casing heads).

Even if a drilling rig is on location, pumping operations usually are done by a service company's cementing unit that can easily mix small batches of cement slurry, measure displacement volume accurately to spot the slurry on bottom, then pump at very low rates and high pressures during the squeeze itself, and finally measure volumes accurately again when reversing out any excess slurry. A squeeze manifold is a compact arrangement of valves and pressure gauges that allows monitoring of the drill pipe and casing pressures throughout the job, and facilitates quick switching of the pumping pressure to either side while the fluid returning from the other side of well is directed to the mud pit or a disposal pit or tank.

The generic term "squeeze" also can apply to injection of generally small volumes of other liquids (e.g., treating fluids) into a zone under pressure. Bullhead squeeze (or just plain bullheading) refers to pumping kill-weight mud down the casing beneath closed blowout preventers in a kick-control situation when it isn't feasible to circulate in such from bottom.

Pumpjack

A pumpjack (also called nodding donkey, pumping unit, horsehead pump, rocking horse, beam pump, dinosaur, sucker rod pump (SRP), grasshopper pump, thirsty bird, or jack pump) is the overground drive for a reciprocating piston pump in an oil well. It is used to mechanically lift liquid out of the well if there is not enough bottom hole pressure for the liquid to flow all the way to the surface. The arrangement is commonly used for onshore wells producing little oil. Pumpjacks are common in oil-rich areas.

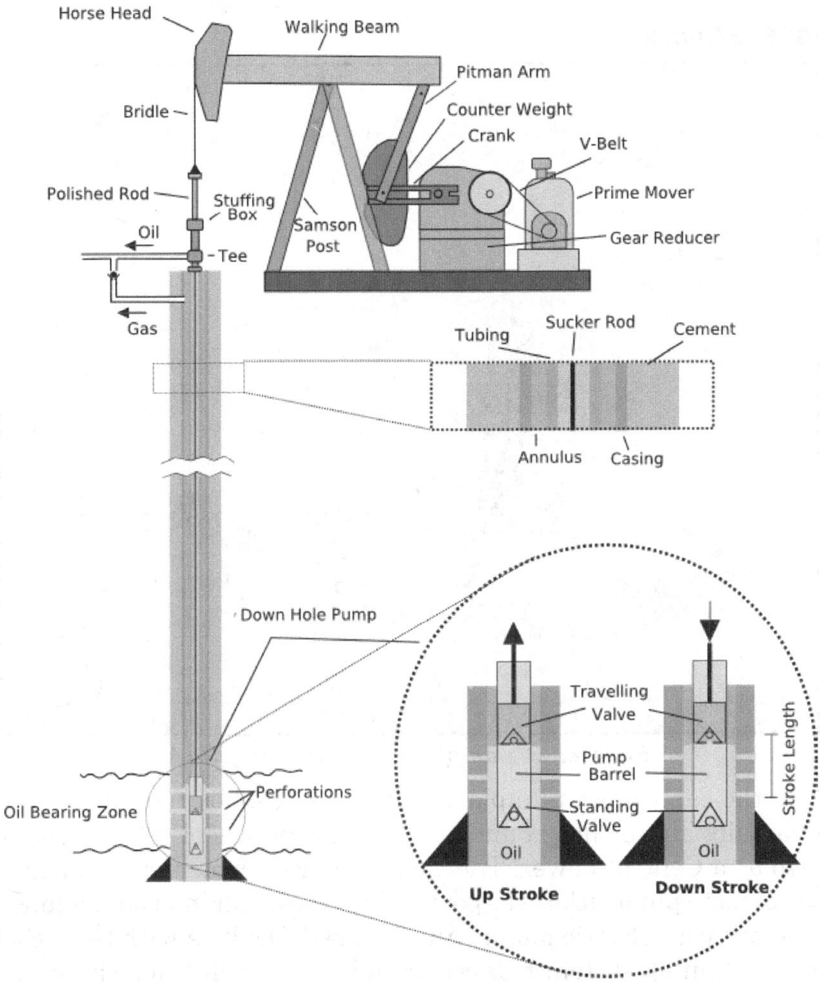

Figure: *A diagram of a pumpjack.*

Depending on the size of the pump, it generally produces 5 to 40 litres of liquid at each stroke. Often this is an emulsion of crude oil and water. Pump size is also determined by the depth and weight of the oil to remove, with deeper extraction requiring more power to move the increased weight of the discharge column (discharge head).

A pumpjack converts the rotary mechanism of the motor to a vertical reciprocating motion to drive the pump shaft, and is exhibited in the characteristic nodding motion. The engineering term for this type of mechanism is a walking beam. It was often employed in stationary and marine steam engine designs in the 18th and 19th centuries.

Above Ground

Figure: *A pumpjack in California, USA*

In the early days, pumpjacks were actuated by rod lines running horizontally above the ground to an eccentric wheel in a mechanism known as a Central Power. The Central Power, which might operate a dozen or more pumpjacks, was powered by a steam or internal combustion engine or by an electric motor. Among the difficulties with this scheme was maintaining system balance as individual well loads changed.

Modern pumpjacks are powered by a prime mover. This is commonly an electric motor, but internal combustion engines are used in isolated locations without access to electricity. Common off-grid pumpjack engines run on casing gas produced from the well, but pumpjacks have been run on many types of fuel, such as propane and Diesel fuel. In harsh climates such motors and engines may be housed in a shack for protection from the elements.

The prime mover of the pumpjack runs a set of pulleys to the transmission which drives a pair of cranks, generally with counterweights on them to assist the motor in lifting the heavy string of rods. The cranks raise and lower one end of an I-beam which is free to move on an A-frame. On the other end of the beam, there is a curved metal box called a horse head or donkey head, so named due to its appearance. A cable made of steel — occasionally, fibreglass — called a bridle, connects the horse head to the polished rod, a piston that passes through the stuffing box.

The polished rod has a close fit to the stuffing box, letting it move in and out of the tubing without fluid escaping. (The tubing is a pipe that runs to the bottom of the well through which the liquid is produced.) The bridle follows the curve of the horse head as it lowers and raises to create a nearly vertical stroke. The polished rod is connected to a long string of rods called sucker rods, which run through the tubing to the down-hole pump, usually positioned near the bottom of the well.

Down-hole

At the bottom of the tubing is the down-hole pump. This pump has two ball check valves: a stationary valve at bottom called the standing valve, and a valve on the piston connected to the bottom of the sucker rods that travels up and down as the rods reciprocate, known as the travelling valve. Reservoir fluid enters from the formation into the bottom of the borehole through perforations that have been made through the casing and cement (the casing is a larger metal pipe that runs the length of the well, which has cement placed between it and the earth; the tubing, pump and sucker rods are all inside the casing).

When the rods at the pump end are travelling up, the travelling valve is closed and the standing valve is open (due to the drop in pressure in the pump barrel). Consequently, the pump barrel fills with the fluid from the formation as the travelling piston lifts the previous contents of the barrel upwards. When the rods begin pushing down, the travelling valve opens and the standing valve closes (due to an increase in pressure in the pump barrel). The travelling valve drops through the fluid in the barrel (which had been sucked in during the upstroke). The piston then reaches the end of its stroke and begins its path upwards again, repeating the process.

Often, gas is produced through the same perforations as the oil. This can be problematic if gas enters the pump, because it can result in what is known as gas locking, where insufficient pressure builds up in the pump barrel to open the valves (due to compression of the gas) and little or nothing is pumped. To preclude this, the inlet for the pump can be placed below the perforations. As the gas-laden fluid enters the well bore through the perforations, the gas bubbles up the annulus (the space between the casing and the tubing) while the liquid moves down to the standing valve inlet. Once at the surface, the gas is collected through piping connected to the annulus.

Water Well Pumpjacks

Pumpjacks can also be used to drive what would now be considered old-fashioned hand-pumped water wells. The scale of the technology is smaller than for an oil well, and can typically fit on top of an existing hand-pumped well head. The technology is simple, typically using a parallel-bar double-cam lift driven from a very low horsepower electric motor. Although the flow rate for a water well pumpjack is low compared to a jet pump and the lifted water is not pressurized, the water well pumpjack has the option of hand pumping in an emergency, by hand-rotating the pumpjack cam to its lowest position, and attaching a manual handle to the top of the wellhead rod.

Artificial Lift

Artificial lift refers to the use of artificial means to increase the flow of liquids, such as crude oil or water, from a production well. Generally this is achieved by the use of a mechanical device inside the well (known as pump or velocity string) or by decreasing the weight of the hydrostatic column by injecting gas into the liquid some distance down the well. Artificial lift is needed in wells when there is insufficient pressure in the reservoir to lift the produced fluids to the surface, but often used in naturally flowing wells (which do not technically need it) to increase the flow rate above what would flow naturally. The produced fluid can be oil, water or a mix of oil and water, typically mixed with some amount of gas.

Usage

Any liquid-producing reservoir will have a 'reservoir pressure': some level of energy or potential that will force fluid (liquid, gas or both) to areas of lower energy or potential. The concept is similar to that of water pressure in a municipal water system. As soon as the pressure inside a production well is decreased below the reservoir pressure, the reservoir will act to fill the well back up, just like opening a valve on a water system. Depending on the depth of the reservoir and density of the fluid, the reservoir may or may not have enough potential to push the fluid to the surface - a deeper well or a heavier mixture results in a higher pressure requirement. Most oil production reservoirs have sufficient potential to naturally produce oil and gas - which are light - during the early phases of production. Water - which is heavier than oil and much heavier than gas - often will eventually encroach into production, possibly causing the well to stop flowing entirely.

Also, reservoir pressure will decrease as many reservoirs deplete, reducing the natural flow to below a profitable rate. At some point, economics can justify the cost of an artificial lift plan to continue or increase production. Most water-producing wells, by contrast, will need artificial lift from the very beginning of production because they do not benefit from the lighter density of oil and gas.

Technologies

Hydraulic Pumping Systems: Hydraulic pumping systems transmit energy to the bottom of the well by means of pressurized power fluid that flows down in the wellbore tubular to a subsurface pump. There are at least three types of hydraulic subsurface pump:

1. a reciprocating piston pump, where one side is powered by the (injected) drive fluid while the other side pumps the produced fluids to surface
2. a jet pump, where the (injected) drive fluid passes through a nozzle-throat venturi combination, mixes with produced fluids and by the venturi effect creates a high pressure at the discharge side of the pump.
3. a hydraulically driven downhole turbine (HSP), whereby the downhole drive motor is a turbine, mechanically connected to the impeller-pump section which pumps the fluid.

These systems are very versatile and have been used in shallow depths (1000 ft) to deeper wells (18,000 ft), low rate wells with production in the tens of barrels per day to wells producing in excess of 20,000 bbl (3,200 m3) per day. In most cases the drive (injected) fluid can be water or produced fluids (oil/water mix). Certain chemicals can be mixed in with the injected fluid to help control corrosion, paraffin and emulsion problems. Hydraulic pumping systems are also suitable for deviated wells where conventional pumps such as the rod pump are not feasible.

Like all systems, these systems have their operating envelopes, though with hydraulic pumps these are often misunderstood by designers. Some types of hydraulic pumps may be sensitive to solids, while jetpumps for example can pump solids volume fractions of more than 50%. They are considered the least efficient lift method, though this differs for the different types of hydraulic pumps, and also when looking at full system losses the differences in many installations are negligible.

The life-cycle cost of these systems is similar to other types of artificial lift when appropriately designed, bearing in mind that they are typically low maintenance, with jet pumps for instance having slightly higher operating (energy) costs with substantially lower purchase cost and virtually no repair cost.

Esp

Electric Submersible Pumps consist of a downhole pump (a series of centrifugal pumps), an electrical motor which transforms the electrical power into kinetic energy to turn the pump, a separator or protector to prevent produced fluids from entering the electrical motor, and an electric power cable that connects the motor to the surface control panel. ESP is a very versatile artificial lift method and can be found in operating environments all over the world. They can handle a very wide range of flow rates (from 200 to 90,000 barrels (14,000 m^3) per day) and lift requirements (from virtually zero to 10,000 ft (3,000 m) of lift). They can be modified to handle contaminants commonly found in oil, aggressive corrosive fluids such as H_2S and CO_2, and exceptionally high downhole temperatures. Increasing water cut has been shown to have no significant detrimental effect on the ESP performance. It is possible to locate them in vertical, deviated, or horizontal wells, but it is recommended to deploy them in a straight section of casing for optimum run life performance.

Although latest developments are aimed to enhance the ESP capabilities to handle gas and sand, they still need more technological development to avoid gas locks and internal erosion. Until recently, ESPs have come with an often prohibitive price tag due to the cost of deployment which can be in excess of $20,000.

Various tools such as Automatic Diverter Valves (ADV), SandCats and other Tubing String and Pump Tools enhance the performance of the ESP. The majority of systems deployed in today's market are Dual ESP Systems which is a simple arrangement of two ESPs in the same well. This delivers a complete downhole system booster or back up - downtime is minimal, workovers cost less and there are savings in other operational areas. ESP Dual Systems bring a significant enhancement of well profitability.

Gas Lift

Gas lift is another widely used artificial lift method. As the name denotes, gas is injected in the tubing to reduce the weight of the hydrostatic column, thus reducing the back pressure and allowing the reservoir pressure to push the mixture of produce fluids and gas up to the surface. The gas lift can be deployed in a wide range of well conditions (from 30,000 bbl/d (4,800 m^3/d) to 15,000 ft (4,600 m)). Gas lifts can cope well with abrasive elements and sand, and the cost of workover is minimum.

Gas lifted wells are equipped with side pocket mandrels and gas lift injection valves. This arrangement allows a deeper gas injection in the tubing. The gas lift system has some disadvantages. There has to be a source of gas, some flow assurance problems such as hydrates can be triggered by the gas lift.

This uses the injection of gas into the fluid stream which reduces the fluid density and lowers the bottom hole pressure. As the gas rises the bubbles help to push the oil ahead. The degree of the effect depends on continuous or intermittent flow of the gas. The gas can be injected at a single point below the fluid or may be supplemented by multipoint injection. An intermitter at the surface controls the timing of the gas injection.

The mechanisms are either pressure or fluid operated. They may be throttling valves or casing pressure operated valve. Fluid operated valves require a rise in tubing pressure to open and drop to close. A throttling pressure valve is opened by casing pressure build up and closed by casing pressure drop. Conventional gas lift valves are attached to gas lift mandrels and wire line retrievable gas lift valves which are set in side pocket mandrels.

PCP

Progressing Cavity Pumps, PCP, are also widely applied in the oil industry. The PCP consists of a stator and a rotor. The rotor is rotated using either a top side motor or a bottom hole motor. The rotation created sequential cavities and the produced fluids are pushed to surface. The PCP is a flexible system with a wide range of applications in terms of rate(up to 5,000 bbl/d (790 m^3/d) and 6,000 ft (1,800 m)). They offer outstanding resistance to abrasives and solids but they are restricted to setting depths and temperatures. Some components of the produced fluids like aromatics can also deteriorate the stator's elastomer.

Rod Pumps

Rod pumps are long slender cylinders with both fixed and moveable elements inside. The pump is designed to be inserted inside the tubing of a well and its main purpose is to gather fluids from beneath it and lift them to the surface. The most important components are: the barrel, valves (travelling and fixed) and the piston. It also has another 18 to 30 components which are called "fittings".

Components

Every part of the pump is important for its correct operation. The most commonly used parts are described below:

- Barrel: The barrel is a long cylinder, which can be from 10 to 36 feet (11 m) long, with a diameter of 1.25 inches (32 mm) to 3.75 inches (95 mm). After experience with several materials for its construction, the API (American Petroleum Institute) standardized the use of two materials or compositions for this part: carbon steel and brass, both with an inside coating of chrome. The advantage of brass against the harder carbon steel is its 100% resistance to corrosion.
- Piston/Plunger: This is a nickel-metal sprayed steel cylinder that goes inside the barrel. Its main purpose is to create a sucking effect that lifts the fluids beneath it and then, with the help of the valves, take the fluids above it, progressively, out of the well. It achieves this with a reciprocating up and down movement.
- Valves: The valves have two components - the seat and the ball - which create a complete seal when closed. The most commonly used seats are made of carbon nitride and the ball is often made of silicon nitride. In the past, balls of iron, ceramic and titanium were used. Titanium balls are still being used but only where crude oil is extremely dense and/or the quantity of fluid to be lifted is large. The most common configuration of a rod pump requires two valves, called the travelling valve and the fixed (or static or standing) valve.
- Piston rod: This is a rod that connects the piston with the outside of the pump. Its main purpose is to transfer the up/down reciprocating energy produced by the "Nodding Donkey" (pumping unit) installed above ground.
- Fittings: The rest of the parts of the pump are called fittings and are, basically, small pieces designed to keep everything

hold together in the right place. Most of these parts are designed to let the fluids pass uninterrupted.

- Filter/Strainer: The job of the filter, as guessed, is to stop big parts of rock, rubber or any other garbage that might be loose in the well from being sucked into the pump. There are several types of filters, with the most common being an iron cylinder with enough holes in it to permit the entrance of the amount of fluid the pump needs.

Sub-Surface Pumping

The sub-surface pump displaces the fluid at the bottom of the well thus lowering the bottom hole pressure. The movement of the plunger and the travelling valve helps to create a low pressure thus moving fluid up the well. The travelling valve is opened on the down stroke and closed on the upstroke. It is on the up stroke that it carries the fluid up the well. The sucker rod is usually 25 ft. long. There are 3 types of pumping units Class 1, Mark 2, or air balanced. By changing the stroke length or the pump rate the production rate can be changed.

The production measured in barrels per day can be calculated with the following formula: P=SxNxC, where P=Production in barrels per day, S=Downhole stroke length (inches), N=Number of strokes per minute, C=A constant derived from the following:

Plunger Diameter = Constant "C" 1 1/16" = 0.132 1 1/4" = 0.182 1 1/2" = 0.262 1 3/4" = 0.357 2" = 0.468 2 1/4" = 0.590 2 1/2" = 0.728 2 3/4" = 0.881 3 1/4" = 1.231 3 3/4" = 1.639

For an Online Calculator: Don-Nan Sucker Rod Pump Production Calculator (bpd)

Production at 100% is theoretical. 80% is a more realistic production calculation.

Rodless Pumping

These can be either hydraulic or electric submersible. The hydraulic uses high pressure power fluid to operate down hole fluid engine. The engine in turn drives a piston that moves the fluid to the surface. The power fluid system can be either open or closed, it depends on whether the power fluid can be mixed with well fluid. This type of system usually has above ground power fluid pumps and a reservoir. The electric submersible is another type of rodless pumping system. This uses an electric pump submerged in the well and connected to a series of transformers and control equipment that power and control the pumping rate. In this system the electric motor is isolated from the oil by a protector. The fluid intake which is before the pump mechanism has a gas separator, also the junction box on the surface helps to dissipate any gas that may have come up the power lines.

Essentially the rod and rodless pumping mechanisms help to achieve the fluid movement by reducing the bottom hole pressure by displacing the fluid above it all by mechanical means. Another method is the plunger lift mechanism which utilizes the tubing string as the barrel. It uses gas to power a plunger.

It is important to note that there are several variations of these methods that can be used. They include; jet pumping involving a hydraulic pump and nozzle that transfers fluid momentum directly to the producing fluid or chamber lift which is a modified gas lift mechanism that has no back pressure. There are also modified rod pumping design units that use either a winch or pneumatic mechanism to work.

Enhanced Oil Recovery

Enhanced Oil Recovery (abbreviated EOR) is a generic term for techniques for increasing the amount of crude oil that can be extracted from an oil field. Enhanced oil recovery is also called improved oil recovery or tertiary recovery (as opposed to primary and secondary recovery). Sometimes the term quaternary recovery is used to refer to more advanced, speculative, EOR techniques. Using EOR, 30 to 60 percent or more of the reservoir's original oil can be extracted, compared with 20 to 40 percent using primary and secondary recovery.

There currently are several different methods of Enhanced Oil Recovery. including steam flood and water flood injection and hydraulic fracturing. Enhanced oil recovery extraction methods consume large quantities of water(Glassman & 2011 4) and natural gas. "Petroleum from the Canadian oil sands extracted via surface mining techniques can consume 20 times more water than conventional oil drilling." However, by 2011 there was inadequate data on the amount of water used in the increasingly important steam-assisted gravity drainage technique (SAGD) method.(Glassman & 2011 4)

Terms

The volumetric sweep efficiency at any time is the fraction of the total reservoir volume contacted by the injected fluid during the recovery. When using water, consideration of the mobility of the fluids, is an important factor when determining the area and vertical sweep efficiencies. This would help to determine the mobility ratio. If M is less than 1 then oil is capable of travelling at a rate equivalent to the water. An increase in the viscosity of the oil would mean that M would increase and this would lead to the injected fluid moving around the oil. This would also make it harder for the oil to penetrate the pore. To improve this ratio then the viscosity of the water has to be increased. When M is greater than 1 the displacing fluid has greater mobility than the displaced fluid. Also the position of the water injection and the flooding patterns would go a long way to determining the recovery patterns. Also to consider in oil recovery is the position and orientation of the injection wells around the production well. As the mobility ratio increases the sweep efficiency decreases. Once a channel of water exists between the injector and the producer then little additional oil would be recovered.

If permeability varies vertically then an irregular vertical fluid front can develop and this is as a result of the differing permeabilities and the mobility ratio. Displacement efficiency refers to the fraction of oil that is swept from unit volume of reservoir upon injection. This depends on the mobility ratio, the wettability of the rock and the pore geometry. The wettability is determined by whether or not the grains preferentially absorb oil over water.

Gas Injection

Gas injection or miscible flooding is presently the most-commonly used approach in enhanced oil recovery. Miscible flooding is a general term for injection processes that introduce miscible gases into the reservoir. A miscible displacement process maintains reservoir pressure and improves oil displacement because the interfacial tension between oil and water is reduced. This refers to removing the interface between the two interacting fluids. This allows for total displacement efficiency.

3E Global, along with its strategic alliances, are working with mineral owners and energy companies to offer comprehensive EOR programmes. Their process involves a proprietary all organic hydro carbon chemical used to infuse oil in the formation with Super Heated Nitrogen and CO2. This process reduces surface tension while raising the gravity of the oil thus increasing production by 100% - 700% or more in most wells. As an added benefit this process often enhances production in surrounding wells.

Gases used include CO_2, natural gas or nitrogen. The fluid most commonly used for miscible displacement is carbon dioxide because it reduces the oil viscosity and is less expensive than liquefied petroleum gas. Oil displacement by carbon dioxide injection relies on the phase behaviour of the mixtures of that gas and the crude, which are strongly dependent on reservoir temperature, pressure and crude oil composition.

Plasma-Pulse

Plasma-Pulse technology is the newest technique used in the USA as of 2013. The technology originated in the Russian Federation at the St. Petersburg State Mining University with funding and assistance from the Skolkovo. The development team in Russia and deployment teams across Russia, Europe and now the USA have experienced this technology in vertical wells with nearly 90% of wells showing positive effects. The technology of plasma-pulse action is absolutely clean, safe, it does not harm the cement column and underground equipment. This oil well technology is protected by multiple international patents.

The Plasma-Pulse Oil Well EOR utilizes low energy emissions to create the same effect that many other technologies can produce except without negative ecological impact. In nearly every case the the volume of water pulled with the oil is actually reduced from pre-EOR treatment instead of increased. Current clients and users of the new technology include ConocoPhillips, ONGC, Gazprom, Rosneft and Lukoil.

It is based in the same technology as the Russian Pulsed Plasma Thruster which was used on two space ships and they are currently advancing the technology for use in horizontal wells.

Miscible Solvents

The injection of various chemicals, usually as dilute solutions, have been used to aid mobility and the reduction in surface tension. Injection of alkaline or caustic solutions into reservoirs with oil that has organic acids naturally occurring in the oil will result in the production of soap that may lower the interfacial tension enough to increase production. Injection of a dilute solution of a water soluble polymer to increase the viscosity of the injected water can increase the amount of oil recovered in some formations. Dilute solutions of surfactants such as petroleum sulfonates or biosurfactants such as rhamnolipids may be injected to lower the interfacial tension or capillary pressure that impedes oil droplets from moving through a reservoir. Special formulations of oil, water and surfactant, microemulsions, can be particularly effective in this. Application of these methods is usually limited by the cost of the chemicals and their adsorption and loss onto the rock of the oil containing formation. In all of these methods the chemicals are injected into several wells and the production occurs in other nearby wells.

Polymer Flooding

Polymer flooding is a means of injecting long chain polymer molecules in an effort to increase the injected water viscosity. The addition of these chemicals means that the fluid would behave like a non-Newtonian fluid; at low velocities it is resistant to flow. This method not only improves the mobility ratio (by lowering it) but also the vertical and areal sweep efficiency. The polymer causes a reduction in the permeability and allows the preferential filling of the high permeable zones in the reservoir. This lowers flow velocity and increases the sweep area.

Surfactant Polymer flooding these are surface active agents that help to break down the surface tension between the oil and water. This allows for the oil and water to separate. The effect of the surfactant depends on the concentration. In low concentrations the rate is gradual but in higher concentrations the rate is increased until such time that the surfactant is diluted by the formation fluids. It also improves the mobility of the fluids and reverses the rock wettability.

Primary surfactants usually have Co-surfactants, activity boosters, Co-solvents added to them to improve stability of the formulation.

Caustic flooding is the addition of sodium hydroxide to injection water to aid recovery. It does this by lowering the surface tension, reversing the rock wettability, emulsification of the oil, mobilization of the oil and helps in drawing the oil out of the rock.

Microbial Injection

Microbial injection is part of microbial enhanced oil recovery and is rarely used because of its higher cost and because the developments is not widely accepted. These microbes function either by partially digesting long hydrocarbon molecules, by generating biosurfactants, or by emitting carbon dioxide (which then functions as described in Gas injection above).

Three approaches have been used to achieve microbial injection. In the first approach, bacterial cultures mixed with a food source (a carbohydrate such as molasses is commonly used) are injected into the oil field. In the second approach, used since 1985, nutrients are injected into the ground to nurture existing microbial bodies; these nutrients cause the bacteria to increase production of the natural surfactants they normally use to metabolize crude oil underground. After the injected nutrients are consumed, the microbes go into near-shutdown mode, their exteriors become hydrophilic, and they migrate to the oil-water interface area, where they cause oil droplets to form from the larger oil mass, making the droplets more likely to migrate to the wellhead. This approach has been used in oilfields near the Four Corners and in the Beverly Hills Oil Field in Beverly Hills, California.

The third approach is used to address the problem of paraffin wax components of the crude oil, which tend to precipitate as the crude flows to the surface, since the Earth's surface is considerably cooler than the petroleum deposits (a temperature drop of 9-10-14 °C per thousand feet of depth is usual).

Liquid Carbon Dioxide Superfluids

Carbon dioxide is particularly effective in reservoirs deeper than 2,000 ft., where CO2 will be in a supercritical state. In high pressure applications with lighter oils, CO_2 is miscible with the oil, with resultant swelling of the oil, and reduction in viscosity, and possibly also with a reduction in the surface tension with the reservoir rock. In the case of low pressure reservoirs or heavy oils, CO_2 will form an immiscible fluid, or will only partially mix with the oil. Some oil swelling may occur, and oil viscosity can still be significantly reduced.

In these applications, between one-half and two-thirds of the injected CO_2 returns with the produced oil and is usually re-injected into the reservoir to minimize operating costs. The remainder is trapped in the oil reservoir by various means. Carbon Dioxide as a solvent has the benefit of being more economical than other similarly miscible fluids such as propane and butane.

Hydrocarbon Displacement

Hydrocarbon displacement is where a slug of hydrocarbon gas is pushed into the reservoir in order to form a miscible phase at high pressure. This however suffers from poor mobility ratio, and the solvent's ability to dissolve the oil is reduced as it goes through. As with all methods, this is only attempted when it is deemed economical.

Thermal Methods

In this approach, various methods are used to heat the crude oil in the formation to reduce its viscosity and/or vaporize part of the oil and thus decrease the mobility ratio. The increased heat reduces the surface tension and increases the permeability of the oil. The heated oil may also vaporize and then condense forming improved oil. Methods include cyclic steam injection, steam drive and combustion. These methods improve the sweep efficiency and the displacement efficiency. Steam injection has been used commercially since the 1960s in California fields. In 2011 solar thermal enhanced oil recovery projects were started in California and Oman, this method is similar to thermal EOR but uses a solar array to produce the steam.

Steam flooding is one means of introducing heat to the reservoir by pumping steam into the well with a pattern similar to that of water injection. Eventually the steam condenses to hot water, in the steam zone the oil evaporates and in the hot water zone the oil expands. As a result the oil expands the viscosity drops and the permeability increases. To ensure success the process has to be cyclical. This is the principal enhanced oil recovery programme in use today.

Fire Flood

Fire flood works best when the oil saturation and porosity are high. Combustion generates the heat within the reservoir itself. Continuous injection of air or other gas mixture with high oxygen content will maintain the flame front. As the fire burns, it moves through the reservoir toward production wells. Heat from the fire reduces oil viscosity and helps vaporize reservoir water to steam. The steam, hot water, combustion gas and a bank of distilled solvent all act to drive oil in front of the fire toward production wells.

There are three methods of combustion: Dry forward, reverse and wet combustion. Dry forward uses an igniter to set fire to the oil. As the fire progresses the oil is pushed away from the fire toward the producing well. In reverse the air injection and the ignition occur from opposite directions. In wet water is injected just behind the front and turned into steam by the hot rock this quenches the fire and spreads the heat more evenly.

Economic Costs and Benefits

Adding oil recovery methods adds to the cost of oil —in the case of CO_2 typically between 0.5-8.0 US$ per tonne of CO_2. The increased extraction of oil on the other hand, is an economic benefit with the revenue depending on prevailing oil prices. Onshore EOR has paid in the range of a net 10-16 US$ per tonne of CO_2 injected for oil prices of 15-20 US$/barrel. Prevailing prices depend on many factors but can determine the economic suitability of any procedure, with more procedures and more expensive procedures being economically viable at higher prices. Example: With oil prices at around 90 US$/barrel, the economic benefit is about 70 US$ per tonne CO_2.

Examples of Current EOR Projects

In Canada, a CO_2-EOR project has been established by Cenovus Energy at the Weyburn Oil Field in southern Saskatchewan since 2000. The project is expected to inject a net 18 million ton CO_2 and recover an additional 130 million barrels (21,000,000 m^3) of oil, extending the life of the oil field by 25 years.(Brown 2001) There is a projected 26+ million tonnes (net of production) of CO2 to be stored in Weyburn, plus another 8.5 million tonnes (net of production) stored at the Weyburn-Midale Carbon Dioxide Project, resulting in a net reduction in atmospheric CO_2). That's the equivalent of taking nearly 7 million cars off the road for a year. Since CO_2 injection began in late 2000, the EOR project has performed largely as predicted. Currently, some 1600 m^3 (10,063 barrels) per day of incremental oil is being produced from the field.

Potential for EOR in United States

The United States has been using EOR for several decades. For over 30 years, oil fields in the Permian Basin have implemented CO2 EOR using naturally sourced CO2 from New Mexico and Colorado. The Department of Energy (DOE) has estimated that full use of 'next generation' CO_2-EOR in United States could generate an additional 240 billion barrels (38 km^3) of recoverable oil resources. Developing this potential would depend on the availability of commercial CO_2 in large volumes, which could be made possible by widespread use of carbon capture and storage.

For comparison, the total undeveloped US domestic oil resources still in the ground total more than 1 trillion barrels (160 km^3), most of it remaining unrecoverable. The DOE estimates that if the EOR potential were to be fully realized, state and local treasuries would gain $280 billion in revenues from future royalties, severance taxes, and state income taxes on oil production, aside from other economic benefits.

Environmental Impacts

Enhanced oil recovery wells typically produce large quantities of brine at the surface. The brine may contain toxic metals and radioactive substances, as well as being very salty. This can be very damaging to drinking water sources and the environment generally if not properly controlled.

In the United States, injection well activity is regulated by the United States Environmental Protection Agency (EPA) and state governments under the Safe Drinking Water Act. EPA has issued Underground Injection Control (UIC) regulations in order to protect drinking water sources. Enhanced oil recovery wells are regulated as Class II wells by the EPA. The regulations require well operators to reinject the brine used for recovery deep underground in Class II Disposal Wells.

Oil and Water Nexus

SAGD, a thermal recovery process, consumes large quantities of water(Glassman & 2011 4) and natural gas. "Petroleum from the Canadian oil sands extracted via surface mining techniques can consume 20 times more water than conventional oil drilling." However, by 2011 there was inadequate data on the amount of water used in the increasingly important steam-assisted gravity drainage technique (SAGD) method.(Glassman & 2011 4)

Hubbert Peak Theory

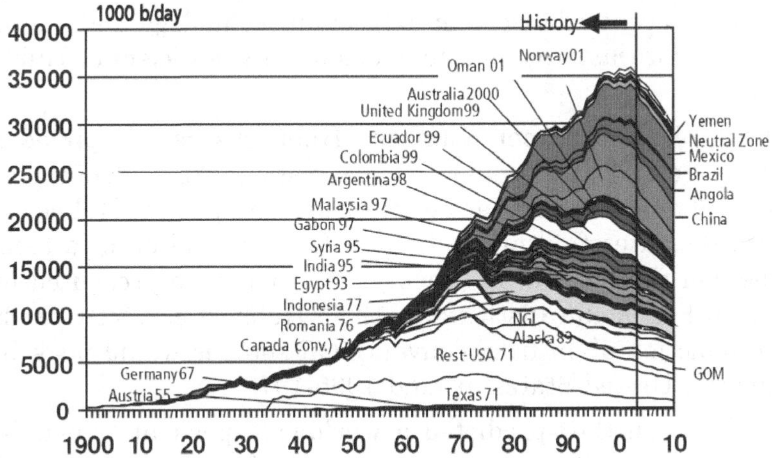

Figure: *2004 U.S. government predictions for oil production other than in OPEC and the former Soviet Union*

The Hubbert peak theory says that for any given geographical area, from an individual oil-producing region to the planet as a whole, the rate of petroleum production tends to follow a bell-shaped curve. It is one of the primary theories on peak oil.

Choosing a particular curve determines a point of maximum production based on discovery rates, production rates and cumulative production. Early in the curve (pre-peak), the production rate increases because of the discovery rate and the addition of infrastructure. Late in the curve (post-peak), production declines because of resource depletion.

The Hubbert peak theory is based on the observation that the amount of oil under the ground in any region is finite, therefore the rate of discovery which initially increases quickly must reach a maximum and decline. In the US, oil extraction followed the discovery curve after a time lag of 32 to 35 years. The theory is named after American geophysicist M. King Hubbert, who created a method of modelling the production curve given an assumed ultimate recovery volume.

Hubbert's Peak

"Hubbert's peak" can refer to the peaking of production of a particular area, which has now been observed for many fields and regions. Hubbert's Peak was thought to have been achieved in the continental US in the early 1970s. Oil production peaked at 10,200,000 barrels per day (1,620,000 m^3/d) and then declined for several years since. Yet, recent advances in extraction technology and optimistic production forecasts has led to widespread skepticism of Hubbert's theory.

Peak oil as a proper noun, or "Hubbert's peak" applied more generally, refers to a singular event in history: the peak of the entire planet's oil production. After Peak Oil, according to the Hubbert Peak Theory, the rate of oil production on Earth would enter a terminal decline. On the basis of his theory, in a paper he presented to the American Petroleum Institute in 1956, Hubbert correctly predicted that production of oil from conventional sources would peak in the continental United States around 1965-1970.

Hubbert further predicted a worldwide peak at "about half a century" from publication and approximately 12 gigabarrels (GB) a year in magnitude. In a 1976 TV interview Hubbert added that the actions of OPEC might flatten the global production curve but this would only delay the peak for perhaps 10 years.

Use of Multiple Curves

The sum of multiple Hubbert curves, a technique not developed by Hubbert himself, may be used in order to model more complicated real life scenarios.

Definition of Reserves

Almost all of Hubbert peaks must be put in the context of high ore grade. Except for fissionable materials, any resource, including oil, is theoretically recoverable from the environment with the right technology. In contrast, Hubbert was concerned with "easy" oil, "easy" metals, and so forth that could be recovered without greatly advanced mining efforts and how to time the necessity of such resource acquisition advancements or substitutions by knowing an "easy" resource's probable peak. Also, as reserves become more difficult to extract there is the possibility that mining or alternatives are too expensive for developing countries.

For heavy crude or deep water drilling attempts, such as Noxal oil field or tar sands or oil shale, the price of the oil extracted will have to include the extra effort required to mine these resources. According to the U.S. Bureau of Ocean Energy Management, Regulation and Enforcement (formerly, the Minerals Management Service), areas such as the Outer Continental Shelf may also incur higher costs due to environmental concerns. So not all oil reserves are equal, and the more difficult reserves are predicted by Hubbert as being typical of the post-peak side of the Hubbert curve.

Reliability

Hubbert, in his 1956 paper, presented two scenarios for US conventional oil production (crude oil + condensate):

- most likely estimate: a logistic curve with a logistic growth rate equal to 6%, an ultimate resource equal to 150 Giga-barrels (Gb) and a peak in 1965.
- upper-bound estimate: a logistic curve with a logistic growth rate equal to 6% and ultimate resource equal to 200 Giga-barrels and a peak in 1970.

Hubbert's upper-bound estimate, which he regarded as optimistic, accurately predicted that US oil production would peak in 1970. Forty years later, the upper-bound estimate has also proven to be very accurate in terms of cumulative production, less so in terms of annual production. For 2005, the upper-bound Hubbert model predicts 178.2 Gb cumulative and 1.17 Gb current production; actual US production was 176.4 Gb cumulative crude oil + condensate (1% lower than the upper bound estimate), with annual production of 1.55 Gb (32% higher than the upper bound estimate).

A post-hoc analysis of peaked oil wells, fields, regions and nations found that Hubbert's model was the "most widely useful" (providing the best fit to the data), though many areas studied had a sharper "peak" than predicted.

Energy Return on Energy Investment

When oil production first began in the mid-nineteenth century, the largest oil fields recovered fifty barrels of oil for every barrel used in the extraction, transportation and refining. This ratio is often referred to as the Energy Return on Energy Investment (EROI or EROEI). Currently, between one and five barrels of oil are recovered for each barrel-equivalent of energy used in the recovery process. As the EROEI drops to one, or equivalently the Net energy gain falls to zero, the oil production is no longer a net energy source. This happens long before the resource is physically exhausted.

Note that it is important to understand the distinction between a barrel of oil, which is a measure of oil, and a barrel of oil equivalent (BOE), which is a measure of energy. Many sources of energy, such as fission, solar, wind, and coal, are not subject to the same near-term supply restrictions that oil is. Accordingly, even an oil source with an EROEI of 0.5 can be usefully exploited if the energy required to produce that oil comes from a cheap and plentiful energy source. Availability of cheap, but hard to transport, natural gas in some oil fields has led to using natural gas to fuel enhanced oil recovery. Similarly, natural gas in huge amounts is used to power most Athabasca Tar Sands plants. Cheap natural gas has also led to Ethanol fuel produced with a net EROEI of less than 1, although figures in this area are controversial because methods to measure EROEI are in debate.

Growth-based Economic Models

Insofar as economic growth is driven by oil consumption growth, post-peak societies must adapt. Hubbert believed:

> *"Our principal constraints are cultural. During the last two centuries we have known nothing but exponential growth and in parallel we have evolved what amounts to an exponential-growth culture, a culture so heavily dependent upon the continuance of exponential growth for its stability that it is incapable of reckoning with problems of nongrowth."*

Some economists describe the problem as uneconomic growth or a false economy. At the political right, Fred Ikle has warned about "conservatives addicted to the Utopia of Perpetual Growth". Brief oil interruptions in 1973 and 1979 markedly slowed - but did not stop - the growth of world GDP.

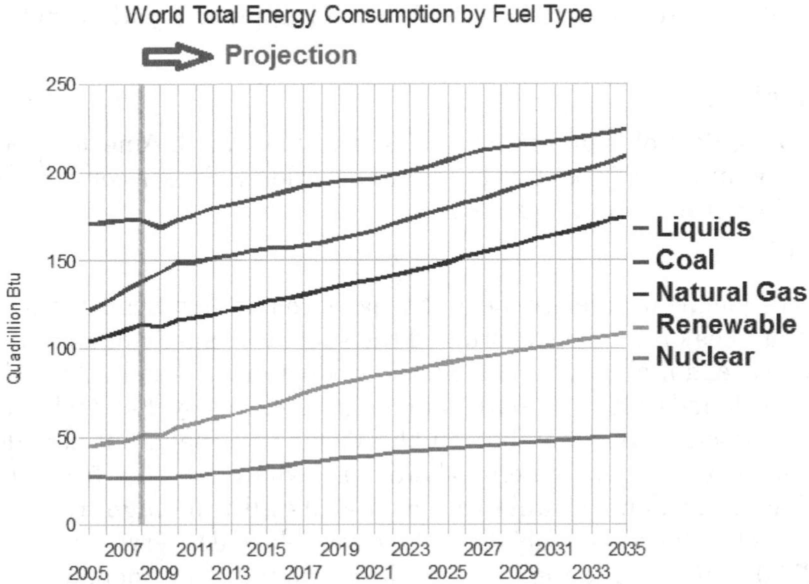

Figure: *World energy consumption & predictions, 2005-2035.* Source: International Energy Outlook 2011.

Between 1950 and 1984, as the Green Revolution transformed agriculture around the globe, world grain production increased by 250%. The energy for the Green Revolution was provided by fossil fuels in the form of fertilizers (natural gas), pesticides (oil), and hydrocarbon fueled irrigation.

David Pimentel, professor of ecology and agriculture at Cornell University, and Mario Giampietro, senior researcher at the National Research Institute on Food and Nutrition (INRAN), place in their study *Food, Land, Population and the U.S. Economy* the maximum U.S. population for a sustainable economy at 200 million. To achieve a sustainable economy world population will have to be reduced by two-thirds, says the study.

Without population reduction, this study predicts an agricultural crisis beginning in 2020, becoming critical c. 2050. The peaking of global oil along with the decline in regional natural gas production may precipitate this agricultural crisis sooner than generally expected. Dale Allen Pfeiffer claims that coming decades could see spiralling food prices without relief and massive starvation on a global level such as never experienced before.

Hubbert Peaks

Although Hubbert peak theory receives most attention in relation to peak oil production, it has also been applied to other natural resources.

Natural Gas

Doug Reynolds predicted in 2005 that the North American peak would occur in 2007. Bentley predicted a world "decline in conventional gas production from about 2020".

Coal

Although observers believe that peak coal is significantly further out than peak oil, Hubbert studied the specific example of anthracite in the USA, a high grade coal, whose production peaked in the 1920s. Hubbert found that anthracite matches a curve closely. Pennsylvania's coal production also matches Hubbert's curve closely, but this does not mean that coal in Pennsylvania is exhausted—far from it. If production in Pennsylvania returned at its all time high, there are reserves for 190 years. Hubbert had recoverable coal reserves worldwide at 2500×10^9 metric tons and peaking around 2150 (depending on usage).

More recent estimates suggest an earlier peak. *Coal: Resources and Future Production* (PDF 630KB), published on April 5, 2007 by the Energy Watch Group (EWG), which reports to the German Parliament, found that global coal production could peak in as few as 15 years. Reporting on this, Richard Heinberg also notes that the date of peak annual energetic extraction from coal is likely to come earlier than the date of peak in quantity of coal (tons per year) extracted as the most energy-dense types of coal have been mined most extensively. A second study, *The Future of Coal* by B. Kavalov and S. D. Peteves of the Institute for Energy (IFE), prepared for European Commission Joint Research Centre, reaches similar conclusions and states that ""coal might not be so abundant, widely available and reliable as an energy source in the future".

Work by David Rutledge of Caltech predicts that the total of world coal production will amount to only about 450 gigatonnes. This implies that coal is running out faster than usually assumed.

Finally, insofar as global peak oil and peak in natural gas are expected anywhere from imminently to within decades at most, any increase in coal production (mining) per annum to compensate for declines in oil or natural gas production, would necessarily translate to an earlier date of peak as compared with peak coal under a scenario in which annual production remains constant.

Fissionable Materials

In a Paper in 1956, after a Review of US Fissionable Reserves, Hubbert Notes of Nuclear Power: There is promise, however, provided mankind can solve its international problems and not destroy itself with nuclear weapons, and provided world population (which is now expanding at such a rate as to double in less than a century) can somehow be brought under control, that we may at last have found an energy supply adequate for our needs for at least the next few centuries of the "foreseeable future."

Technologies such as the thorium fuel cycle, reprocessing and fast breeders can, in theory, considerably extend the life of uranium reserves. Roscoe Bartlett claims

Our current throwaway nuclear cycle uses up the world reserve of low-cost uranium in about 20 years.

Caltech physics professor David Goodstein has stated that

... you would have to build 10,000 of the largest power plants that are feasible by engineering standards in order to replace the 10 terawatts of fossil fuel we're burning today ... that's a staggering amount and if you did that, the known reserves of uranium would last for 10 to 20 years at that burn rate. So, it's at best a bridging technology ... You can use the rest of the uranium to breed plutonium 239 then we'd have at least 100 times as much fuel to use. But that means you're making plutonium, which is an extremely dangerous thing to do in the dangerous world that we live in.

Helium

Almost all helium on Earth is a result of radioactive decay of uranium and thorium. Helium is extracted by fractional distillation from natural gas, which contains up to 7% helium. The world's largest helium-rich natural gas fields are found in the United States, especially in the Hugoton and nearby gas fields in Kansas, Oklahoma, and Texas. The extracted helium is stored underground in the National Helium Reserve near Amarillo, Texas, the self-proclaimed "Helium Capital of the World". Helium production is expected to decline along with natural gas production in these areas.

Helium is the second-lightest chemical element in the Universe, causing it to rise to the upper layers of Earth's atmosphere. Helium atoms are so light that the Earth's gravity field is simply not strong enough to trap helium in the atmosphere and it dissipates slowly into space and is lost forever.

Transition Metals

Hubbert applied his theory to "rock containing an abnormally high concentration of a given metal" and reasoned that the peak production for metals such as copper, tin, lead, zinc and others would occur in the time frame of decades and iron in the time frame of two centuries like coal. The price of copper rose 500% between 2003 and 2007 and was attributed by some to peak copper. Copper prices later fell, along with many other commodities and stock prices, as demand shrank from fear of a global recession. Lithium availability is a concern for a fleet of Li-ion battery using cars but a paper published in 1996 estimated that world reserves are adequate for at least 50 years. A similar prediction for platinum use in fuel cells notes that the metal could be easily recycled.

Precious Metals

The possibility of Peak Gold has emerged recently [5]. Aaron Regent president of the Canadian gold giant Barrik Gold said that global output has been falling by roughly one million ounces a year since the start of the decade. The total global mine supply has dropped by 10pc as ore quality erodes, implying that the roaring bull market of the last eight years may have further to run. "There is a strong case to be made that we are already at 'peak gold'," he told The Daily Telegraph at the RBC's annual gold conference in London. "Production peaked around 2000 and it has been in decline ever since, and we forecast that decline to continue. It is increasingly difficult to find ore," he said.

Ore grades have fallen from around 12 grams per tonne in 1950 to nearer 3 grams in the US, Canada, and Australia. South Africa's output has halved since peaking in 1970. Output fell a further 14 percent in South Africa in 2008 as companies were forced to dig ever deeper - at greater cost - to replace depleted reserves.

Phosphorus

Phosphorus supplies are essential to farming and depletion of reserves is estimated at somewhere from 60 to 130 years. According to a 2008 study, the total reserves of phosphorus are estimated to be approximately 3,200 MT, with a peak production at 28 MT/year in 2034. Individual countries' supplies vary widely; without a recycling initiative America's supply is estimated around 30 years. Phosphorus supplies affect agricultural output which in turn limits alternative fuels such as biodiesel and ethanol. Its increasing price and scarcity (global price of rock phosphate rose 8-fold in the 2 years to mid 2008) could change global agricultural patterns. Lands, perceived as marginal because of remoteness, but with very high phosphorus content, such as the Gran Chaco may get more agricultural development, while other farming areas, where nutrients are a constraint, may drop below the line of profitability.

Peak Water

Hubbert's original analysis did not apply to renewable resources. However, over-exploitation often results in a Hubbert peak nonetheless. A modified Hubbert curve applies to any resource that can be harvested faster than it can be replaced.

For example, a reserve such as the Ogallala Aquifer can be mined at a rate that far exceeds replenishment. This turns much of the world's underground water and lakes into finite resources with peak usage debates similar to oil. These debates usually centre around agriculture and suburban water usage but generation of electricity from nuclear energy or coal and tar sands mining mentioned above is also water resource intensive. The term fossil water is sometimes used to describe aquifers whose water is not being recharged.

Renewable Resources

- Fisheries: At least one researcher has attempted to perform Hubbert linearization (Hubbert curve) on the whaling industry, as well as charting the transparently dependent price of caviar on sturgeon depletion. Another example is the cod of the North Sea. The comparison of the cases of fisheries and of mineral extraction tells us that the human pressure on the environment is causing a wide range of resources to go through a depletion cycle which follows a Hubbert curve.

Criticisms of Peak Oil

Economist Michael Lynch argues that the theory behind the Hubbert curve is too simplistic and relies on an overly Malthusian point of view. Lynch claims that Campbell's predictions for world oil production are strongly biased towards underestimates, and that Campbell has repeatedly pushed back the date.

Leonardo Maugeri, vice president of the Italian energy company Eni, argues that nearly all of peak estimates do not take into account unconventional oil even though the availability of these resources is significant and the costs of extraction and processing, while still very high, are falling because of improved technology. He also notes that the recovery rate from existing world oil fields has increased from about 22% in 1980 to 35% today because of new technology and predicts this trend will continue. The ratio between proven oil reserves and current production has constantly improved, passing from 20 years in 1948 to 35 years in 1972 and reaching about 40 years in 2003. These improvements occurred even with low investment in new exploration and upgrading technology because of the low oil prices during the last 20 years. However, Maugeri feels that encouraging more exploration will require relatively high oil prices.

Edward Luttwak, an economist and historian, claims that unrest in countries such as Russia, Iran and Iraq has led to a massive underestimate of oil reserves. The Association for the Study of Peak Oil and Gas (ASPO) responds by claiming neither Russia nor Iran are troubled by unrest currently, but Iraq is.

Cambridge Energy Research Associates authored a report that is critical of Hubbert-influenced predictions:

Despite his valuable contribution, M. King Hubbert's methodology falls down because it does not consider likely resource growth, application of new technology, basic commercial factors, or the impact of geopolitics on production. His approach does not work in all cases-including

on the United States itself-and cannot reliably model a global production outlook. Put more simply, the case for the imminent peak is flawed. As it is, production in 2005 in the Lower 48 in the United States was 66 percent higher than Hubbert projected.

CERA does not believe there will be an endless abundance of oil, but instead believes that global production will eventually follow an "undulating plateau" for one or more decades before declining slowly, and that production will reach 40 Mb/d by 2015.

Alfred J. Cavallo, while predicting a conventional oil supply shortage by no later than 2015, does not think Hubbert's peak is the correct theory to apply to world production.

Chapter 8

Drilling Technologies

The latest study of Massachusetts Institute of Technology "The Future of Geothermal Energy – Impact of Enhanced Geothermal Systems (EGS) on the United States in the 21st Century" (2006) points out the essential importance of developing an economical deep geothermal boring technology. With current boring technologies, bore price rises exponentially with depth. Thus, finding a boring technology with which the bore price rise would be approximately linear with increasing bore depth is an important challenge.

New Drilling Technologies Requirements

This MIT study characterizes the requirements on new fast and ultra-deep boring technology as follows:

- the price of boring rises linearly with depth
- bore axis with neutral floating
- the possibility to make vertical or inclined bores up to 10 km deep
- the possibility to make large diameter bores – even 5 times larger than on the ground compared to today drilling technologies
- casing formed on site in the borehole

Examples of New Drilling Technologies

There are more than 20 research efforts solving innovative drilling technology such as: laser, spallation, plasma, electron beam, pallets, enhanced rotary, electric spark and discharge, electric arc, water jet erosion, ultrasonic, chemical, induction, nuclear, forced flame explosive, turbine, high frequency, microwave, heating/cooling stress, electric current and several other. The most promising solutions are mentioned below:

1. Hydrothermal spallation – Thermal spallation drilling uses a large, downhole burner, much like a jet engine, to apply a high

heat flux to the rock face. This drilling technology is based on thermal processes of rock spallation and fusion.

2. Chemical plasma – is based on crushing by high-speed combustion, but nitric acid as oxidizing agent instead of oxygen.
3. Erosion - most patents refer to water jet rock cutting. Different modification variants are described, e.g. utilization of cavitation, turbulent processes, combination with mechanical processes, etc.
4. Laser - during the recent decade intense research has been made into utilization of high energy laser beams for rock disintegration. Primarily conversion of military equipment is concerned. Laser energy is used for the process of thermal spallation, melting, or evaporation of rock.
5. Electric discharge - The methods utilizing electric discharge are based on long-term experience gained in other application areas.
6. Electrical plasma - is based on crushing by irradiation of plasma with high temperature up to 20 000°C
7. Direct transfer of heat - This technology is based on electrically melting rock at 1400°C; lava gravel will float to top; bore hole walls are of glass of surrounding rock. Cost decreases with depth, with no limit on depth of bore hole. Bore diameters from 1m to 10m. Recovery of energy used to melt rock.

High-energetic Electrical Plasma

One of the most promising approaches in deep drilling field is utilization of electrical plasma. It has lower energy efficiency than some of the other technologies, but it has several other advantages. Producing boreholes with wide range of diameters or drilling in water environment can be mentioned. The research team from Slovakia has developed drilling concept based on utilization of electrical plasma. The core of the research is held in Research Centre for Deep Drilling which was opened in the premises of Slovak Academy of Sciences in Bratislava.

State of the Art

No one from these until now proved to be effective in severe conditions and no one is solving the problem in complex including the energy and material transport from 5 to 10 km technically and economically. Therefore, intensive support is required in research phase of these technologies.

Plasma Deep Drilling Technology

Plasma deep drilling technology is one of several different variants of recently explored new drilling technologies which would be able to substitute conventional, contact-based rotary systems. Worldwide there are several research teams which work continually on research and development of new deep drilling technology. In addition to plasma, other concepts have been based on different principles such as water jet, hydrothermal spallation or laser. Until now, none of these other attempts has been proven to be effective in extreme deep drilling conditions.

High-energetic Electrical Plasma

High-energetic electrical plasma is a promising technology which is currently being developed in deep drilling applications. It has lower energy efficiency than some of the other technologies, but has several other advantages. The most important are utilization in water environment or producing boreholes with wide range of diameters. The research team from Slovakia is developing drilling concept based on electrical plasma principle suitable for several drilling applications. The core of the research is held in Research Centre for Deep Drilling established in the premises of Slovak Academy of Sciences in Bratislava.

Physical Principle of Electrical Plasma

An electric arc is an electrical breakdown of a gas which produces an ongoing plasma discharge, resulting from a current flowing through normally nonconductive media such as air or gas. An arc discharge is characterized by a lower voltage than a glow discharge, and relies on thermionic emission of electrons from the electrodes supporting the arc. The electric arc is influenced by different factors: the gas flow, inner and outer magnetic fields, construction elements of the chamber which confine the arc, the elements being often under electric voltage, differing from that of the arc. Thus, there is a strong need for deep understanding of a wide spectrum of the processes taking place in the discharge chamber in order to develop highly effective plasma torches that we have proposed to use as a source of the thermal plasma.

Plasma Deep Drilling Technology in Mining Applications

Plasma deep drilling technology may be suitable not only for geothermal, oil and gas or carbon capture and storage applications but also for underground mining. This new non-contact drilling method combined with the development of current mining techniques may bring smart technology for underground mining industry. The most promising project connects advantages of plasma drilling and raise boring technologies.

Advantages of Plasma Deep Drilling Technology

1. Higher drilling energy efficiency
2. Continuous drilling process without replacement of mechanical parts
3. Constant casing diameter
4. Effective transport of disintegrated rock

Drilling Rig

A drilling rig is a machine which creates holes in the ground. Drilling rigs can be massive structures housing equipment used to drill water wells, oil wells, or natural gas extraction wells, or they can be small enough to be moved manually by one person and are called auger. They sample sub-surface mineral deposits, test rock, soil and groundwater physical properties, and also can be used to install sub-surface fabrications, such as underground utilities, instrumentation, tunnels or wells. Drilling rigs can be mobile equipment mounted on trucks, tracks or trailers, or more permanent land or marine-based structures (such as oil platforms, commonly called 'offshore oil rigs' even if they don't contain a drilling rig). The term "rig" therefore generally refers to the complex of equipment that is used to penetrate the surface of the Earth's crust.

Drilling rigs can be:

- Small and portable, such as those used in mineral exploration drilling, water wells and environmental investigations. Energold's frontier S-Style drills for example focus on portability, while conventional water well rigs are far heavier for long-term drilling projects
- Huge, capable of drilling through thousands of metres of the Earth's crust. Large "mud pumps" circulate drilling mud (slurry) through the drill bit and up the casing annulus, for cooling and removing the "cuttings" while a well is drilled. Hoists in the rig can lift hundreds of tons of pipe. Other equipment can force acid or sand into reservoirs to facilitate extraction of the oil or natural gas; and in remote locations there can be permanent living accommodation and catering for crews (which may be more than a hundred). Marine rigs may operate many hundreds of miles or kilometres distant from the supply base with infrequent crew rotation or cycle.

Petroleum Drilling Industry

Figure: *Drilling the Bakken formation in the Williston Basin*

Figure: *Swanson Drilling drilling directional oil wells in East Texas*

Oil and natural gas drilling rigs can be used not only to identify geologic reservoirs but also to create holes that allow the extraction of oil or natural gas from those reservoirs. Primarily in onshore oil and gas fields once a well has been drilled, the drilling rig will be moved off of the well and a service rig (a smaller rig) that is purpose-built for completions will be moved on to the well to get the well on line. This frees up the drilling rig to drill another hole and streamlines the operation as well as allowing for specialization of certain services, i.e., completions vs. drilling.

Water Well Drilling

New portable drillcat technology uses smaller portable trailer mounted rigs with shorter 10 foot (3.0 m) drill pipe. DIY users and missionary groups and even preppers use these to drill water wells as they can be operated by 1 or 2 people with a minimal skill level. The shorter drill pipe also allows a much smaller mast, which lets you have a much smaller and lighter rig which is cheaper to ship overseas and can fit in a standard 20 foot (6.1 m) shipping container. Portable trailer mounted drilling rigs have drill ratings from 300 to 800 feet (91 to 240 m) depending on mud pump flow and pressure ratings and drill pipe sizes.

Figure: *Rig drilling popularly known as 'Borewell' in operation for Water well drilling, in India.*

Other, heavier, truck rigs are more complicated, thus requiring more skill to run. They're also more difficult to handle safely due to the longer 20 to 30 foot (6.1 to 9.1 m) drill pipe. Large truck rigs also require a much higher overhead clearance to operate. Large truck drills can use over 150 US gallons (570 L) or more of fuel per day, while the smaller Deeprock Style portable drills use a mere 5 to 20 US gallons (19 to 76 L) of fuel per day. The most desired portable rigs are USA made, because of the Steel quality. This makes smaller, more portable rigs preferable in remote or hard-to-reach places, and are more cost effective in this new era of high fuel prices.

History

Figure: *Antique drilling rig now on display at Western History Museum in Lingle, Wyoming. It was used to drill many water wells in that area—many of those wells are still in use.*

Figure: *Antique drilling Rigs in Zigong, China*

Until internal combustion engines were developed in the late 19th century, the main method for drilling rock was muscle power of man or animal. The drilling of wells for the manufacture of salt began by the Song Dynasty in China. The well had a particularly small mouth, "as small as a small bowl." Archaeological evidence of the drilling tools used in deep-well dwelling are kept and displayed in the Zigong Salt Industry Museum. According to *Salt: A World History*, a Qing Dynasty well, also located in Zigong, "continued down to 3,300 feet making it at the time the deepest drilled well in the world." Mechanised versions of this system persisted until about 1970, using a cam to rapidly raise and drop what, by then, was a steel cable up to 3 mm

Figure: *Swanson Drilling Texas East at night*

In the 1970s, outside of the oil and gas industry, roller bits using mud circulation were replaced by the first pneumatic reciprocating piston Reverse Circulation (RC) drills, and became essentially obsolete for most shallow drilling, and are now only used in certain situations where rocks preclude other methods. RC drilling proved much faster and more efficient, and continues to improve with better metallurgy, deriving harder, more durable bits, and compressors delivering higher air pressures at higher volumes, enabling deeper and faster penetration. Diamond drilling has remained essentially unchanged since its inception.

Figure: *Night view of the H&P drilling the Bakken*

Mobile Drilling Rigs

In early oil exploration, drilling rigs were semi-permanent in nature and the derricks were often built on site and left in place after the completion of the well. In more recent times drilling rigs are expensive custom-built machines that can be moved from well to well. Some light duty drilling rigs are like a mobile crane and are more usually used to drill water wells. Larger land rigs must be broken apart into sections and loads to move to a new place, a process which can often take weeks.

Small mobile drilling rigs are also used to drill or bore piles. Rigs can range from 100 ton continuous flight auger (CFA) rigs to small air powered rigs used to drill holes in quarries, etc. These rigs use the same technology and equipment as the oil drilling rigs, just on a smaller scale.

The drilling mechanisms outlined below differ mechanically in terms of the machinery used, but also in terms of the method by which drill cuttings are removed from the cutting face of the drill and returned to surface.

Drilling Rig Classification

There are many types and designs of drilling rigs, with many drilling rigs capable of switching or combining different drilling technologies as needed. Drilling rigs can be described using any of the following attributes:

By Power Used

- Mechanical — the rig uses torque converters, clutches, and transmissions powered by its own engines, often diesel
- Electric — the major items of machinery are driven by electric motors, usually with power generated on-site using internal combustion engines
- Hydraulic — the rig primarily uses hydraulic power
- Pneumatic — the rig is primarily powered by pressurized air
- Steam — the rig uses steam-powered engines and pumps (obsolete after middle of 20th Century)

By Pipe Used

- Cable — a cable is used to raise and drop the drill bit
- Conventional — uses metal or plastic drill pipe of varying types
- Coil tubing — uses a giant coil of tube and a downhole drilling motor

By Height

(Rigs are differentiated by height based on how many connected pipe they are able to "stand" in the derrick when needing to temporarily remove the drill pipe from the hole. Typically this is done when changing a drill bit or when "logging" the well.)

- Single — can pull only single drill pipes. The presence or absence of vertical pipe racking "fingers" varies from rig to rig.
- Double — can hold a stand of pipe in the derrick consisting of two connected drill pipes, called a "double stand".

- Triple — can hold a stand of pipe in the derrick consisting of three connected drill pipes, called a “triple stand”.
- Quadri — can store stand of pipe in the derrick composed of four connected drill pipes, called a “quadri stand”.

By Method of Rotation or Drilling Method

- No-rotation includes direct push rigs and most service rigs
- Rotary table — rotation is achieved by turning a square or hexagonal pipe (the “Kelly”) at drill floor level.
- Top drive — rotation and circulation is done at the top of the drill string, on a motor that moves in a track along the derrick.
- Sonic — uses primarily vibratory energy to advance the drill string
- Hammer — uses rotation and percussive force

By Position of Derrick

- Conventional — derrick is vertical
- Slant — derrick is slanted at a 45 degree angle to facilitate horizontal drilling

Drill Types

There are a variety of drill mechanisms which can be used to sink a borehole into the ground. Each has its advantages and disadvantages, in terms of the depth to which it can drill, the type of sample returned, the costs involved and penetration rates achieved. There are two basic types of drills: drills which produce rock chips, and drills which produce core samples.

Auger Drilling

Auger drilling is done with a helical screw which is driven into the ground with rotation; the earth is lifted up the borehole by the blade of the screw. Hollow stem auger drilling is used for softer ground such as swamps where the hole will not stay open by itself for environmental drilling, geotechnical drilling, soil engineering and geochemistry reconnaissance work in exploration for mineral deposits. Solid flight augers/bucket augers are used in harder ground construction drilling. In some cases, mine shafts are dug with auger drills. Small augers can be mounted on the back of a utility truck, with large augers used for sinking piles for bridge foundations.

Auger drilling is restricted to generally soft unconsolidated material or weak weathered rock. It is cheap and fast.

Figure: *Cable tool water well drilling rig in Kimball, West Virginia. These slow rigs have mostly been replaced by rotary drilling rigs in the U.S.*

Percussion Rotary air Blast Drilling (RAB)

RAB drilling is used most frequently in the mineral exploration industry. (This tool is also known as a Down-the-hole drill.) The drill uses a pneumatic reciprocating piston-driven "hammer" to energetically drive a heavy drill bit into the rock. The drill bit is hollow, solid steel and has ~20 mm thick tungsten rods protruding from the steel matrix as "buttons". The tungsten buttons are the cutting face of the bit.

The cuttings are blown up the outside of the rods and collected at surface. Air or a combination of air and foam lift the cuttings.

RAB drilling is used primarily for mineral exploration, water bore drilling and blast-hole drilling in mines, as well as for other applications such as engineering, etc. RAB produces lower quality samples because the cuttings are blown up the outside of the rods and can be contaminated from contact with other rocks. RAB drilling at extreme depth, if it encounters water, may rapidly clog the outside of the hole with debris, precluding removal of drill cuttings from the hole. This can be counteracted, however, with the use of "stabilizers" also known as "reamers", which are large cylindrical pieces of steel attached to the drill string, and made to perfectly fit the size of the hole being drilled. These have sets of rollers on the side, usually with tungsten buttons, that constantly break down cuttings being pushed upwards.

The use of high-powered air compressors, which push 900-1150 cfm of air at 300-350 psi down the hole also ensures drilling of a deeper hole up to ~1250 m due to higher air pressure which pushes all rock cuttings and any water to the surface. This, of course, is all dependent on the density and weight of the rock being drilled, and on how worn the drill bit is.

Air Core Drilling

Air core drilling and related methods use hardened steel or tungsten blades to bore a hole into unconsolidated ground. The drill bit has three blades arranged around the bit head, which cut the unconsolidated ground. The rods are hollow and contain an inner tube which sits inside the hollow outer rod barrel. The drill cuttings are removed by injection of compressed air into the hole via the annular area between the innertube and the drill rod. The cuttings are then blown back to surface up the inner tube where they pass through the sample separating system and are collected if needed. Drilling continues with the addition of rods to the top of the drill string. Air core drilling can occasionally produce small chunks of cored rock.

This method of drilling is used to drill the weathered regolith, as the drill rig and steel or tungsten blades cannot penetrate fresh rock. Where possible, air core drilling is preferred over RAB drilling as it provides a more representative sample. Air core drilling can achieve depths approaching 300 metres in good conditions. As the cuttings are removed inside the rods and are less prone to contamination compared to conventional drilling where the cuttings pass to the surface via outside return between the outside of the drill rod and the walls of the hole. This method is more costly and slower than RAB.

Cable Tool Drilling

Cable tool rigs are a traditional way of drilling water wells. The majority of large diameter water supply wells, especially deep wells completed in bedrock aquifers, were completed using this drilling method. Although this drilling method has largely been supplanted in recent years by other, faster drilling techniques, it is still the most practicable drilling method for large diameter, deep bedrock wells, and in widespread use for small rural water supply wells. The impact of the drill bit fractures the rock and in many shale rock situations increases the water flow into a well over rotary.

Figure: *SpeedStar cable tool drilling rig, Ballston Spa, New York*

Also known as ballistic well drilling and sometimes called "spudders", these rigs raise and drop a drill string with a heavy carbide tipped drilling bit that chisels through the rock by finely pulverizing the subsurface materials. The drill string is composed of the upper drill rods, a set of "jars" (inter-locking "sliders" that help transmit additional energy to the drill bit and assist in removing the bit if it is stuck) and the drill bit. During the drilling process, the drill string is periodically removed from the borehole and a bailer is lowered to collect the drill cuttings (rock fragments, soil, etc.). The bailer is a bucket-like tool with a trapdoor in the base. If the borehole is dry, water is added so that the drill cuttings will flow into the bailer. When lifted, the trapdoor closes and the cuttings are then raised and removed. Since the drill string must be raised and lowered to advance the boring, the casing (larger diameter outer piping) is typically used to hold back upper soil materials and stabilize the borehole.

Cable tool rigs are simpler and cheaper than similarly sized rotary rigs, although loud and very slow to operate. The world record cable tool well was drilled in New York to a depth of almost 12,000 feet (3,700 m). The common Bucyrus Erie 22 can drill down to about 1,100 feet (340 m). Since cable tool drilling does not use air to eject the drilling chips like a rotary, instead using a cable strung bailer, technically there is no limitation on depth.

Cable tool rigs now are nearly obsolete in the United States. They are mostly used in Africa or Third-World countries. Being slow, cable tool rig drilling means increased wages for drillers. In the United States drilling wages would average around US$200 per day per man, while in Africa it is only US$6 per day per man, so a slow drilling machine can still be used in undeveloped countries with depressed wages. A cable tool rig can drill 25 feet (7.6 m) to 60 feet (18 m) of hard rock a day. A newer rotary drillcat top head rig equipped with down-the-hole (DTH) hammer can drill 500 feet (150 m) or more per day, depending on size and formation hardness.

Reverse Circulation (RC) Drilling

Figure: *Reverse Circulation (RC) rig, outside Newman, Western Australia*

RC drilling is similar to air core drilling, in that the drill cuttings are returned to surface inside the rods. The drilling mechanism is a pneumatic reciprocating piston known as a "hammer" driving a tungsten-steel drill bit. RC drilling utilises much larger rigs and machinery and depths of up to 500 metres are routinely achieved. RC drilling ideally produces dry rock chips, as large air compressors dry the rock out ahead of the advancing drill bit. RC drilling is slower and costlier but achieves better penetration than RAB or air core drilling; it is cheaper than diamond coring and is thus preferred for most mineral exploration work.

Figure: *Track mounted Reverse Circulation rig (side view).*

Reverse circulation is achieved by blowing air down the rods, the differential pressure creating air lift of the water and cuttings up the “inner tube”, which is inside each rod. It reaches the “divertor” at the top of the hole, then moves through a sample hose which is attached to the top of the “cyclone”. The drill cuttings travel around the inside of the cyclone until they fall through an opening at the bottom and are collected in a sample bag.

The most commonly used RC drill bits are 5-8 inches (13–20 cm) in diameter and have round tungsten ‘buttons’ that protrude from the bit, which are required to drill through shale and abrasive rock. As the buttons wear down, drilling becomes slower and the rod string can potentially become bogged in the hole. This is a problem as trying to recover the rods may take hours and in some cases weeks. The rods and drill bits themselves are very expensive, often resulting in great cost to drilling companies when equipment is lost down the bore hole.

Most companies will regularly re-grind the buttons on their drill bits in order to prevent this, and to speed up progress. Usually, when something is lost (breaks off) in the hole, it is not the drill string, but rather from the bit, hammer, or stabiliser to the bottom of the drill string (bit). This is usually caused operator error, over-stressed metal, or adverse drilling conditions causing downhole equipment to get stuck in a part of the hole.

Although RC drilling is air-powered, (water is also used, to reduce dust, keep the drill bit cool, and assist in pushing cutting back upwards, but also when "collaring" a new hole. A mud called "Liqui-Pol" is mixed with water and pumped into the rod string, down the hole. This helps to bring up the sample to the surface by making the sand stick together. Occasionally, "Super-Foam" (a.k.a. "Quik-Foam") is also used, to bring all the very fine cuttings to the surface, and to clean the hole. When the drill reaches hard rock, a "collar" is put down the hole around the rods, which is normally PVC piping. Occasionally the collar may be made from metal casing. Collaring a hole is needed to stop the walls from caving in and bogging the rod string at the top of the hole. Collars may be up to 60 metres deep, depending on the ground, although if drilling through hard rock a collar may not be necessary.

Reverse circulation rig setups usually consist of a support vehicle, an auxiliary vehicle, as well as the rig itself. The support vehicle, normally a truck, holds diesel and water tanks for resupplying the rig. It also holds other supplies needed for maintenance on the rig. The auxiliary is a vehicle, carrying an auxiliary engine and a booster engine. These engines are connected to the rig by high pressure air hoses. Although RC rigs have their own booster and compressor to generate air pressure, extra power is needed which usually isn't supplied by the rig due to lack of space for these large engines. Instead, the engines are mounted on the auxiliary vehicle. Compressors on an RC rig have an output of around 1000 cfm at 500 psi (500 $L \cdot s^{-1}$ at 3.4 MPa). Alternatively, stand-alone air compressors which have an output of 900-1150cfm at 300-350 psi each are used in sets of 2, 3, or 4, which are all routed to the rig through a multi-valve manifold.

Diamond Core Drilling

Diamond core drilling (exploration diamond drilling) utilizes an annular diamond-impregnated drill bit attached to the end of hollow drill rods to cut a cylindrical core of solid rock. The diamonds used to make diamond core bits are a variety of sizes, fine to microfine industrial grade diamonds, and the ratio of diamonds to metal used in the matrix affects the performance of the bits cutting ability in different types of rock formations . The diamonds are set within a matrix of varying hardness, from brass to high-grade steel. Matrix hardness, diamond size and dosing can be varied according to the rock which must be cut. The bits made with hard steel with a low diamond count and are ideal for softer highly fractured rock while others made of softer steels and high dimond ratio are good for coring in hard solid rock. Holes within the bit allow water to be delivered to the cutting face. This provides three essential functions — lubrication, cooling, and removal of drill cuttings from the hole.

Figure: *Multi-combination drilling rig (capable of both diamond and reverse circulation drilling). Rig is currently set up for diamond drilling.*

Diamond drilling is much slower than reverse circulation (RC) drilling due to the hardness of the ground being drilled. Drilling of

1200 to 1800 metres is common and at these depths, ground is mainly hard rock. Techniques vary among drill operators and what the rig they are using is capable of, some diamond rigs need to drill slowly to lengthen the life of drill bits and rods, which are very expensive and time consuming to replace at extremely deep depths. As a diamond drill rig cores deeper and deeper the time consuming part of the process is not cutting 5 to 10 more feet of rock core but the retrieval of the core with the whire line & overshot tool. Core samples are retrieved via the use of a core tube, a hollow tube placed inside the rod string and pumped with water until it locks into the core barrel. As the core is drilled, the core barrel slides over the core as it is cut. An "overshot" attached to the end of the winch cable is lowered inside the rod string and locks on to the backend (aka head assembly), located on the top end of the core barrel. The winch is retracted, pulling the core tube to the surface. The core does not drop out of the inside of the core tube when lifted because either a split ring core lifter or basket retainer allow the core to move into, but not back out of the tube.

Figure: *Diamond core drill bits*

Once the core tube is removed from the hole, the core sample is then removed from the core tube and catalogued. The Driller's assistant unscrews the backend off the core tube using tube wrenches, then each part of the tube is taken and the core is shaken out into core trays. The core is washed, measured and broken into smaller pieces using a hammer or sawn through to make it fit into the sample trays. Once catalogued, the core trays are retrieved by geologists who then analyse the core and determine if the drill site is a good location to expand future mining operations.

Diamond rigs can also be part of a multi-combination rig. Multi-combination rigs are a dual setup rig capable of operating in either a reverse circulation (RC) and diamond drilling role (though not at the same time). This is a common scenario where exploration drilling is being performed in a very isolated location. The rig is first set up to drill as an RC rig and once the desired metres are drilled, the rig is set up for diamond drilling. This way the deeper metres of the hole can be drilled without moving the rig and waiting for a diamond rig to set up on the pad.

Direct Push Rigs

Direct push technology includes several types of drilling rigs and drilling equipment which advances a drill string by pushing or hammering without rotating the drill string. While this does not meet the proper definition of drilling, it does achieve the same result — a borehole. Direct push rigs include both cone penetration testing (CPT) rigs and direct push sampling rigs such as a PowerProbe or Geoprobe. Direct push rigs typically are limited to drilling in unconsolidated soil materials and very soft rock.

CPT rigs advance specialized testing equipment (such as electronic cones), and soil samplers using large hydraulic rams. Most CPT rigs are heavily ballasted (20 metric tons is typical) as a counter force against the pushing force of the hydraulic rams which are often rated up to 20 kN. Alternatively, small, light CPT rigs and offshore CPT rigs will use anchors such as screwed-in ground anchors to create the reactive force. In ideal conditions, CPT rigs can achieve production rates of up to 250–300 metres per day.

Direct push drilling rigs use hydraulic cylinders and a hydraulic hammer in advancing a hollow core sampler to gather soil and groundwater samples. The speed and depth of penetration is largely dependent on the soil type, the size of the sampler, and the weight and power of the rig. Direct push techniques are generally limited to shallow soil sample recovery in unconsolidated soil materials. The advantage of direct push technology is that in the right soil type it can produce a large number of high quality samples quickly and cheaply, generally from 50 to 75 metres per day. Rather than hammering, direct push can also be combined with sonic (vibratory) methods to increase drill efficiency.

Hydraulic Rotary Drilling

Oil well drilling utilises tri-cone roller, carbide embedded, fixed-cutter diamond, or diamond-impregnated drill bits to wear away at the cutting face. This is preferred because there is no need to return intact samples to surface for assay as the objective is to reach a formation containing oil or natural gas. Sizable machinery is used, enabling depths of several kilometres to be penetrated. Rotating hollow drill pipes carry down bentonite and barite infused drilling muds to lubricate, cool, and clean the drilling bit, control downhole pressures, stabilize the wall of the borehole and remove drill cuttings. The mud travels back to the surface around the outside of the drill pipe, called the annulus. Examining rock chips extracted from the mud is known as mud logging. Another form of well logging is electronic and is frequently employed to evaluate the existence of possible oil and gas deposits in the borehole. This can take place while the well is being drilled, using Measurement While Drilling tools, or after drilling, by lowering measurement tools into the newly drilled hole.

The rotary system of drilling was in general use in Texas in the early 1900s. It is a modification of one invented by Fauvelle in 1845, and used in the early years of the oil industry in some of the oil-producing countries in Europe. Originally pressurized water was used instead of mud, and was almost useless in hard rock before the diamond cutting bit. The main breakthrough for rotary drilling came in 1901, when Anthony Francis Lucas combined the use of a steam-driven rig and of mud instead of water in the Spindletop discovery well.

The drilling and production of oil and gas can pose a safety risk and a hazard to the environment from the ignition of the entrained gas causing dangerous fires and also from the risk of oil leakage polluting water, land and groundwater. For these reasons, redundant safety systems and highly trained personnel are required by law in all countries with significant production.

Sonic (Vibratory) Drilling

A sonic drill head works by sending high frequency resonant vibrations down the drill string to the drill bit, while the operator controls these frequencies to suit the specific conditions of the soil/rock geology. Vibrations may also be generated within the drill head. The frequency is generally between 50 and 120 hertz (cycles per second) and can be varied by the operator.

Resonance magnifies the amplitude of the drill bit, which fluidizes the soil particles at the bit face, allowing for fast and easy penetration through most geological formations. An internal spring system isolates these vibrational forces from the rest of the drill rig.

Limits of the Technology

Figure: *An oil rig*

Drill technology has advanced steadily since the 19th century. However, there are several basic limiting factors which will determine the depth to which a bore hole can be sunk.

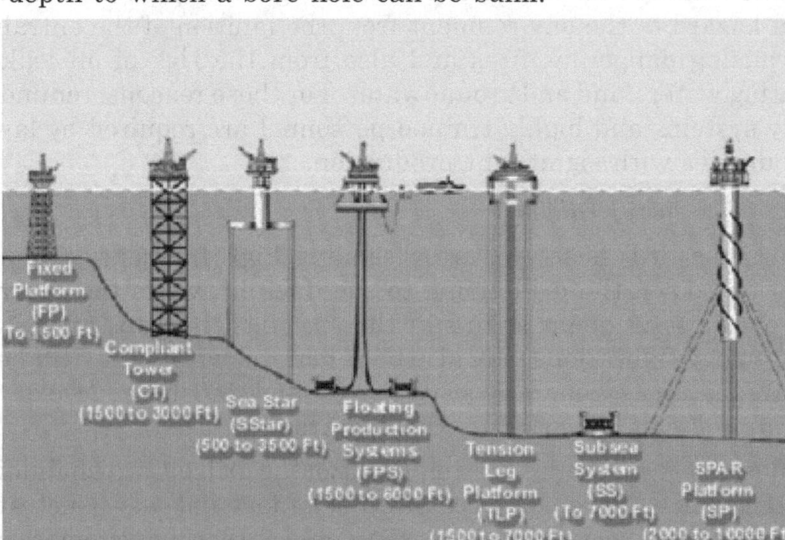

Figure: *Different Drilling Rigs*

All holes must maintain outer diameter; the diameter of the hole must remain wider than the diameter of the rods or the rods cannot turn in the hole and progress cannot continue. Friction caused by the drilling operation will tend to reduce the outside diameter of the drill bit. This applies to all drilling methods, except that in diamond core drilling the use of thinner rods and casing may permit the hole to continue. Casing is simply a hollow sheath which protects the hole against collapse during drilling, and is made of metal or PVC. Often diamond holes will start off at a large diameter and when outside diameter is lost, thinner rods put down inside casing to continue, until finally the hole becomes too narrow. Alternatively, the hole can be reamed; this is the usual practice in oil well drilling where the hole size is maintained down to the next casing point.

For percussion techniques, the main limitation is air pressure. Air must be delivered to the piston at sufficient pressure to activate the reciprocating action, and in turn drive the head into the rock with sufficient strength to fracture and pulverise it. With depth, volume is added to the in-rod string, requiring larger compressors to achieve operational pressures. Secondly, groundwater is ubiquitous, and increases in pressure with depth in the ground. The air inside the rod string must be pressurised enough to overcome this water pressure at the bit face. Then, the air must be able to carry the rock fragments to surface. This is why depths in excess of 500 m for reverse circulation drilling are rarely achieved, because the cost is prohibitive and approaches the threshold at which diamond core drilling is more economic.

Diamond drilling can routinely achieve depths in excess of 1200 m. In cases where money is no issue, extreme depths have been achieved, because there is no requirement to overcome water pressure. However, water circulation must be maintained to return the drill cuttings to surface, and more importantly to maintain cooling and lubrication of the cutting surface of the bit; while at the same time reduce friction on the steel walls of the rods turning against the rock walls of the hole. When water return is lost the rods will vibrate, this is called "rod chatter", and that will damage the drill rods, crack the joints.

Without sufficient lubrication and cooling, the matrix of the drill bit will soften. While diamond is the hardest substance known, at 10 on the Mohs hardness scale, it must remain firmly in the matrix to achieve cutting. Weight on bit, the force exerted on the cutting face of the bit by the drill rods in the hole above the bit, must also be monitored.

A unique drilling operation in deep ocean water was named Project Mohole.

Research of New Drilling Technologies

Research includes technologies based on the utilization of water jet, chemical plasma, hydrothermal spallation or laser.

Utilization of high energetic electrical plasma shows promise in deep drilling applications. This approach has potential to replace conventional drilling technologies. It would be able to produce boreholes with large constant diameter without frequent replacement of the drill bits.

Causes of Deviation

Most drill holes deviate slightly from their planned trajectory. This is because of the torque of the turning bit working against the cutting face, because of the flexibility of the steel rods and especially the screw joints, because of reaction to foliation and structure within the rock, and because of refraction as the bit moves into different rock layers of varying resistance. Additionally, inclined holes will tend to deviate upwards because the drill rods will lie against the bottom of the bore, causing the drill bit to be slightly inclined from true. It is because of deviation that drill holes must be surveyed if deviation will impact the usefulness of the information returned. Sometimes the surface location can be offset laterally to take advantage of the expected deviation tendency, so the bottom of the hole will end up near the desired location. Oil well drilling commonly uses a process of controlled deviation called directional drilling (e.g., when several wells are drilled from one surface location).

Rig Equipment

Drilling rigs typically include at least some of the following items:

- Blowout preventers: (BOPs)

The equipment associated with a rig is to some extent dependent on the type of rig but (#23 & #24) are devices installed at the wellhead to prevent fluids and gases from unintentionally escaping from the borehole. #23 is the annular (often referred to as the "Hydril", which is one manufacturer) and #24 is the pipe rams and blind rams. In the place of #24 Variable bore rams or VBRs can be used. These offer the same pressure and sealing capacity found in standard pipe rams, while offering the versatility of sealing on various sizes of drill pipe, production tubing and casing without changing standard pipe rams.

Normally VBRs are used when utilizing a tapered drill string (when different size drill pipe is used in the complete drill string).

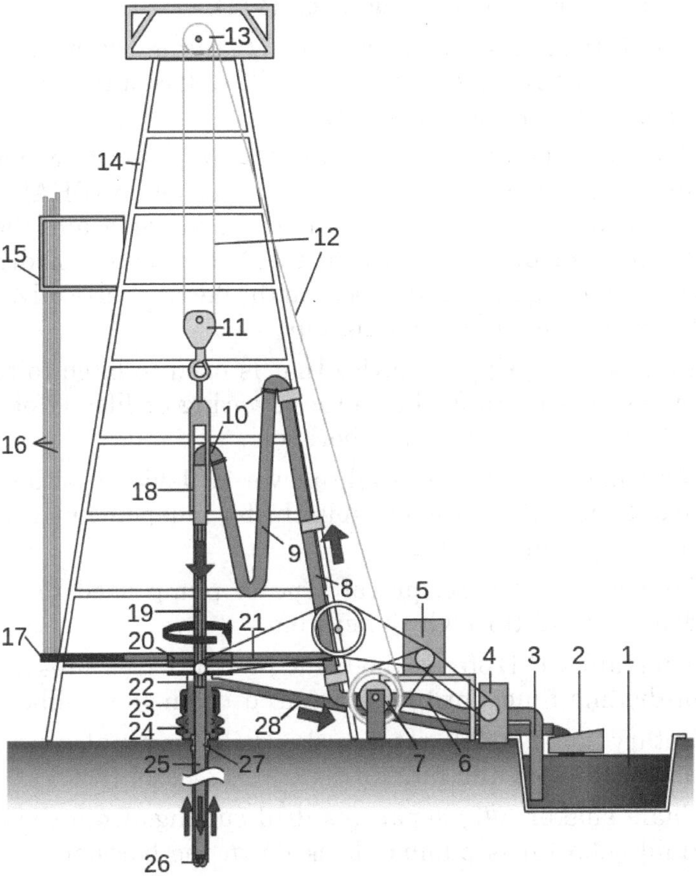

***Figure:** Simple diagram of a drilling rig and its basic operation*

- Centrifuge: an industrial version of the device that separates fine silt and sand from the drilling fluid.
- Solids control: solids control equipments for preparing drilling mud for the drilling rig.
- Chain tongs: wrench with a section of chain, that wraps around whatever is being tightened or loosened. Similar to a pipe wrench.
- Degasser: a device that separates air and/or gas from the drilling fluid.
- Desander / desilter: contains a set of hydrocyclones that separate sand and silt from the drilling fluid.

- Drawworks: (#7) is the mechanical section that contains the spool, whose main function is to reel in/out the drill line to raise/lower the travelling block (#11).
- Drill Bit: (#26) is a device attached to the end of the drill string that breaks apart the rock being drilled. It contains jets through which the drilling fluid exits.
- Drill pipe: (#16) joints of hollow tubing used to connect the surface equipment to the bottom hole assembly (BHA) and acts as a conduit for the drilling fluid. In the diagram, these are "stands" of drill pipe which are 2 or 3 joints of drill pipe connected together and "stood" in the derrick vertically, usually to save time while tripping pipe.
- Elevators: a gripping device that is used to latch to the drill pipe or casing to facilitate the lowering or lifting (of pipe or casing) into or out of the borehole.
- Mud motor: a hydraulically powered device positioned just above the drill bit used to spin the bit independently from the rest of the drill string.
- Mud pump: (#4) reciprocal type of pump used to circulate drilling fluid through the system.
- Mud tanks: (#1) often called mud pits, provides a reserve store of drilling fluid until it is required down the wellbore.
- Rotary table: (#20) rotates the drill string along with the attached tools and bit.
- Shale shaker: (#2) separates drill cuttings from the drilling fluid before it is pumped back down the borehole.

Bibliography

Aghar, Hani; M. Caire, H. Elshahawi, J. R. Gomez, J. Saeedi, C. Young, B. Pinguet, K. Swainson, E. Takla, B. Theuveny: *The Expanding Scope of Well Testing*, Spring, NY, 2007.

Akers, A.; M. Gassman, & R. Smith: *Hydraulic Power System Analysis*, Taylor & Francis, New York, 2006.

Arthur W. McCray and Frank W. Cole: *Oil Well Drilling Technology*, New India Pub, Delhi, 2005.

Brantly, J.E.: *History of Oil Well Drilling,* Gulf Publishing Co., Houston, 1971.

Chen and Lui: *Handbook of Structural Engineering*, CRC Press, NY, 2005.

Chowdhary, L.R. : *Petroleum Geology of the Cambay Basin, Gujarat, India*, Indian Petroleum, Delhi, 2004.

Critchfield, J. W.: *Tunneling in Gassy Ground,* Proceedings, RETC, AIME, New York, 1985.

DeLaisi, Francis: *Oil, Its Influence on Politics.* Labor Publishing,1923.

Derrick A, Francis C, Bokalders, V.: *Solar Photovoltaic Products: A Guide for Development Workers,* Intermediate Technology Publications, London, 1991.

Engler, Robert: *The Politics of Oil: A Study of Private Power and Democratic Directions.* New York, Macmillan, 1961.

Fischer, Louis: *Oil Imperialism: The International Struggle for Petroleum.* New York: International Publishers, 1926.

Franceys R, Pickford J, Reed, R.: *A Guide to the Development of On-site Sanitation*, World Health Organization, Geneva, 1992.

Frankel, Paul: *Mattei: Oil and Power Politics.* New York and Washington: Praeger, 1966.

Gramling, Robert: *Oil on the Edge: Offshore Development, Conflict, Gridlock.* Albany: State University of New York Press, 1996.

Hailu, S.: *Applicability of Small-bore Gravity Sewers in Addis Ababa,* Tampere University of Technology, Ethiopia, Finland, 1988.

Hansmire, W. and Monsees, J.E.: *Evolving Practices for Tunnel Contracts in the United States*, Elsevier, London, 2006,

Hartshorn, J. E.: *Oil Companies and Governments: An Account of the International Oil Industry in Its Political Environment.* London: Faber and Faber, 1962.

Inversin, AR: *Micro-hydropower Sourcebook: a Practical Guide to Design and Implementation in Developing Countries*, NRECA International Foundation, Washington, DC, 1986.

Jentleson, Bruce: *Pipeline Politics: The Complex Political Economy of East-West Energy Trade.* lthaca: Cornell University Press, 1986.

Kern, C.: *Community Water Development*, Intermediate Technology Publications, London, 1989.

Kuesel, T.: *Immersed Tube Tunnels: Concept Design and Construction*, Boston Society of Civil Engineers, Spring, 1986.

Lancashire S, Kenna J, Fraenkel, P.: *Windpumping Handbook,* Intermediate Technology Publications, London, 1987.

Macaulay, David R. and Jason F. McLennan: *The Ecological Engineer*, Ecotone Publishing, Kansas City, 2006.

McNeil, Ian: *An Encyclopedia of the History of Technology*, Routledge, London, 1990.

McNeil, Ian: *Hydraulic Power*, Longman Group, 1972.

Morgan, P.: *Rural Water Supplies and Sanitation*, Macmillan, London, 1990.

Nilsson, A.: *Groundwater Dams for Small-scale Water supply*, Intermediate Technology Publications, London, 1988.

Olien, Diana Davids: *Oil in Texas: The Gusher Age, 1895–1945.* Austin: University of Texas Press, 2002.

Paretsky, Sara: *Tunnel Vision*, Dell Publishing, New York, 1994.

Phien-Wej, N., and EJ. Cording: *Sheared Shale Response to Deep TBM Excavation*, Elsevier, New York, 1991.

Pollak, F.: *Pump Users Handbook,* Trade & Technical Press, Morden, England, 1988.

Ripley, B. D: *Design of Pressure Tunnels and Shafts*, J. A. Hudson, 1993.

Satyanarayana, D. : *Petroleum Geochemistry*, Daya Publishing, Delhi, 2011.

Schneider, Steven A.: *The Oil Price Revolution.* Baltimore, 1983.

Teng, W.C.: *Foundation Design of Drilling*, Prentice-Hall, Inc., Englewood, NJ, 1962.

Tucker, L.S.: *Sewage Flow Variations in Individual Homes*, American Society of Civil Engineers, 1967.

William H. Rae, Jr.: *Low-Speed Wind Tunnel Testing*, John Wiley & Sons, Inc., New York, 1999.

Winblad, U., Kilama, W.: *Sanitation without Water*, Macmillan, Basingstoke, 1985.

Wood, A.M.: *Tunneling, Management,* E & FN Spon, London, 2000.

Index

❑❑❑